Probability

Probability

Peter Whittle

Churchill Professor of the Mathematics of
Operational Research, University of Cambridge.
Fellow of Churchill College, Cambridge

JOHN WILEY & SONS

London · New York · Sydney · Toronto

Copyright © P. Whittle 1970
First published by Penguin Books Ltd. 1970
Published by John Wiley & Sons Ltd. 1976

Library of Congress Cataloging in Publication Data:

Whittle, Peter.
  Probability.

  Bibliography: p.
  Includes index.
  1. Probabilities.   I. Title.
QA273.W59   1976   519.2   76-863

ISBN 0 471 01657 8

Reproduced and printed by photolithography and bound in
Great Britain at The Pitman Press, Bath

To my parents

Contents

Preface

This book is intended as a first text in the theory and applications of probability, demanding a reasonable, but not extensive, knowledge of mathematics. It takes the reader to what one might describe as a good intermediate level.

With so many excellent texts available, the provision of another needs justification. One minor motive for my writing this book was the feeling that the very success of certain applications of probability theory in the past has brought about a rather stereotyped treatment of applications in most general texts. I have therefore made an effort to present a wider variety of important applications; for instance: optimization problems, quantum mechanics, information theory and statistical mechanics.

However, the principal novelty of the present treatment is that the theory is based on an axiomatization of the concept of expectation, rather than of that of a probability measure. Such an approach is now preferred in advanced theories of integration and probability: it is interesting that the recent texts of Krickeberg (1965), Neveu (1964) and Feller (1966) all devote some attention to it, although without taking it up whole-heartedly. However, I belive that such an approach has great advantages even in an introductory treatment. There is no great point in arguing the matter: only the text itself can provide real justification. However, I can briefly indicate the reasons for my belief. (i) To begin with, people probably have a better intuitive feeling for what is meant by an 'average value' than for what is meant by a 'probability'. (ii) Certain important topics, such as optimization and approximation problems can then be introduced and treated very quickly, just because they are phrased in terms of expectations. (iii) Most elementary treatments are bedevilled by the apparent need to ring the changes of a particular proof or discussion for all the special cases of continuous or discrete distribution, scalar or vector variables, etc. In the expectation approach these are indeed seen as special cases which can be treated with uniformity and economy. (iv) The operation approach – analysis of the type of assertion that is really relevant in a particular application – leads one surprisingly often to a formulation in expectations. (v) There are advantages at the advanced level, but here we do not need to make a case.

The mathematical demands made upon the reader scarcely go beyond simple analysis and a few basic properties of matrices. Some properties of convex functions or sets are required occasionally – these are explained – and the spectral resolution of a matrix is used in section 12.1. Because of the approach taken, no measure theory is demanded of the reader, and any set theory needed is explained. Probability generating functions and the like are used freely, from an early stage. I feel this to be right for the subject and a manner of thinking that initially requires practice rather than any extensive mathematics. Fourier arguments are confined almost entirely to section 11.2.

It is not necessary to read the whole text. A reader interested principally in applications could restrict himself to chapters 1 and 2, sections 1, 2 and 4 of chapter 3, chapter 4, chapter 5 (except section 4), chapter 6 (except section 2), chapters 7, 8 and 12. If following the theory alone, one could omit sections 3–6 of chapter 6, and chapters 7 and 12.

This project was begun at the University of Manchester, for which I have a lasting respect and affection. It was completed during tenure of my current post, endowed by the Esso Petroleum Company Ltd.

P. WHITTLE

1 Introduction

1.1 A few ideas and examples

Probability is an everyday notion. This is shown by the number of common words related to the idea : chance, random, hazard, fortune, likelihood, odds, uncertainty, expect, believe. Nevertheless, as with many concepts for which we have a strong but rather vague intuition, the idea of probability has taken some time to formalize. It was in the study of games of chance (such as card games and the tossing of dice) that the early attempts at formalization were made, with the understandable motive of determining one's chances of winning. In these cases the basis of formalization was fairly clear, because one can always work from the fact that there are a number of elementary situations (such as all the possible deals in a card game) which can be regarded as 'equally likely'. However, this approach fails in the less tangible problem of physical, economic and human life which we should like to consider, and of which we shall give a few examples below.

As is fortunately common, the absence of a firm basis did not prevent the subject of probability from flourishing greatly in the eighteenth and nineteenth centuries. However, in 1933 a satisfactory general basis was achieved by A. N. Kolmogorov in the form of an axiomatic theory which, while not necessarily the last word on the subject, set the pattern of any future theory.

The application of probability theory to games of chance is an obvious one. However, there are applications in science and technology which are almost as clean cut. As examples, we can quote the genetic mechanism of Mendelian inheritance (sections 5.6 and 7.2), the operation of a telephone exchange (exercise 7.7.2), or the decay of radioactive molecules (section 7.6). In all these cases one can make valuable progress with a simple probability model, although it is only fair to add that a deeper study will demand something more complicated.

In general, the physical sciences constitute a rich source of well-defined and interesting probability problems; see, for example, some of the models of statistical mechanics (section 6.5) and polymer chemistry (exercise 7.6.6). In encountering the 'natural variability' of biological problems one runs into rather more diffuse situations (for example, the analysis of agricultural experiments), but this variability makes a probabilistic approach all the more imperative, and one can construct idealized models of, say, population growth (sections 7.6 and 7.7) which have proved valuable. One meets natural variability in the human form if one tries to construct social or economic models (see section 2.5, for example), but again such models, inevitably probabilistic, prove useful.

One of the most recent and fascinating applications of probability theory is to the field of control or, more generally, to that of sequential decision making (section 12.3). One might, for example, be wishing to hold a rocket on course, despite the fact that random forces of one sort or another tend to divert it, or one might wish to keep a factory in a state of over-all efficient production, despite the fact that so many future variables (such as demand for the product) must be uncertain. In either case, one must make a sequence of decisions (regarding course adjustment or factory management) in such a way as to ensure efficient running over a period, or even optimal running, in some well-defined sense. Moreover, these decisions must be taken in the face of an uncertain future.

It should also be said that probability theory has its own flavour and intrinsic structure, quite apart from applications, as may be apparent from chapters 2, 3, 9 and 10, in particular. Just as for mathematics in general, people argue about the extent to which the theory is self-generating, or dependent upon applications to suggest the right direction and concepts. Perhaps either extreme view is incorrect : the search for an inner pattern and the search for a physical pattern are both powerful research tools, neither of them to be neglected.

1.2   The empirical basis

Certain experiments are non-reproducible in that, when repeated under standard conditions, they produce variable results. The classic example is coin tossing : the toss being the experiment, resulting in the observation of a head or a tail. To take something less artificial, one might be observing the response of a rat to a certain drug, observation on another rat constituting repetition of the experiment. However uniform in quality the experimental animals may be, one will certainly observe a variable response. The same variability would be found in, for example, failure times of electric lamps, crop yields, the collisions of physical particles, or the number of telephone calls made on a weekday in May. This variability cannot always be dismissed as 'experimental error' which could presumably be explained and reduced, but may be something more fundamental. For instance, the ejection of a particular electron from a hot metal filament is a definite event, which may or may not occur, and is not predictable on any physical theory yet developed.

Probability theory can be regarded as an attempt to provide a quantitative basis for the discussion of such situations, or at least for some of them. One might despair of constructing a theory for phenomena whose essential quality is that of imprecision, but there is an empirical observation which gives the needed starting point.

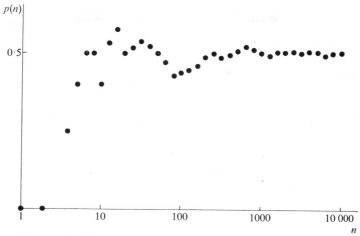

Figure 1 A graph of the proportion of heads thrown, $p(n)$, in a sequence of n throws, from an actual coin-tossing experiment. Note the logarithmic scale for n. The figures are taken from Kerrich (1946), by courtesy of Professor Kerrich and his publishers

Suppose one tosses a coin repeatedly, keeping a record of the number of heads $r(n)$ in the first n tosses ($n = 1, 2, 3, \ldots$). Consider now the proportion of heads after n tosses:

$$p(n) = \frac{r(n)}{n}. \qquad \qquad \textbf{1.2.1}$$

It is an empirical fact that $p(n)$ varies with n much as in Figure 1, which is a genuine graph from such a coin-tossing experiment. The values of $p(n)$ show fluctuations, which become progressively weaker as n increases, until ultimately $p(n)$ shows signs of tending to some kind of limit value, interpretable as the 'long-run proportion' of heads. Obviously this cannot be a limit in the usual mathematical sense because one cannot guarantee that the fluctuations in $p(n)$ will have fallen below a definite level for all values of n from a certain point onwards. However, some kind of 'limit' there seems to be, and it is this fact that offers the hope of a useful theory: that beyond the short-term irregularity there is a long-term regularity.

This same regularity manifests itself if we examine, for example, lifetimes of electric lamp bulbs. Let the observed lifetimes be denoted $X_1, X_2, X_3, \ldots$, and suppose we keep a running record of the arithmetic average of the first n lifetimes:

$$\overline{X}_n = \frac{1}{n} \sum_1^n X_j. \qquad \qquad \textbf{1.2.2}$$

Then, again, it is an empirical fact that, provided we keep test conditions constant, the graph of $\overline{X}_n$ against n will show a similar convergence. That is, fluctuations slowly die down with increasing n, and $\overline{X}_n$ appears to tend to a 'limit' value, interpretable as the 'long-run average' of lifetime for this particular model of bulb.

It is on this feature that one founds probability theory: by postulating the existence of an idealized 'long-run proportion' (a *probability*) or 'long-run average' (an *expectation*).

Actually, the case of a proportion is a special case of that of an average. Suppose that in the coin-tossing experiment one defined an 'indicator variable' X_j which took the value 1 or 0 according as the jth toss resulted in a head or a tail. Then the average (expression **1.2.2**) of X_j would reduce just to the proportion given by equation **1.2.1**. Conversely, one can build up an average from proportions (see exercise 1.2.2).

Correspondingly, in constructing an axiomatic theory one has the choice of two methods: to idealize the concept of a proportion or that of an average. In the first case one starts with the idea of a probability and later builds up that of an expectation. In the second, one takes an expectation as the basic concept, of which a probability is to be regarded as a special case. In this text we shall take the second course, which, although less usual, offers substantial advantages.

Exercises

1.2.1   Carry out your own coin-tossing experiment, and graph the results, as in Figure 1.

1.2.2   Suppose that the observations X_j can only take a discrete set of values

$$X = x_k \quad (k = 1, 2, \ldots).$$

Note that the average $\overline{X}_n$ of formula **1.2.2** can be written

$$\overline{X}_n = \sum_k x_k p_k(n),$$

where $p_k(n)$ is the proportion of times the value x_k is observed in the first n readings.

Readers might feel that our 'non-reproducible experiments' could often be made reproducible if only they were sufficiently refined. For example, in the coin-tossing experiment, the path of the coin is surely mechanically determined, and if the conditions of tossing were standardized sufficiently, then a standard (and theoretically predictable) result should be obtained. For the rat experiment, increased standardization should again produce increased reproducibility. More than this; a sufficiently good understanding of the biological response of rats to the drug should enable one to *predict* the response of a given animal to a given dose and so make the experiment reproducible in the sense that the observed variation could be largely explained.

Whether it is possible in principle (it certainly is not in practice) to remove all variability from an experiment in this way is the philosophic issue of determinism which, if decidable at all, is certainly not decidable in naïve terms. In probability theory we shall simply start from the premise that there is a certain amount of variability which we cannot explain but must accept.

The practical point of coin or die tossing is that the coin or die acts as a 'variability amplifier': the dynamics are such that a small variability in initial position is transformed into a relatively large variability in final position.

A final point of terminology: because probability theory has its historical roots in gaming rather than in science, what we have referred to as an 'experiment' or 'observation' is often termed a 'trial' (a translation of the French *essai*). We shall use the three terms interchangeably, choosing that which suits the context best.

1.3   Averages over a finite population

In this section we shall make our closest acquaintance with official statistics.

Suppose that a country of n people has a 100 per cent census and that, as a result of the information gathered, people can be assigned to one of K mutually exclusive categories or cells which we shall label $\omega_1, \omega_2, \ldots, \omega_K$. Thus the specification characterizing a person in a particular category might conceivably be: 'born in the United Kingdom in 1935, male, married, three children, motor mechanic, with an income of between £1600 and £1700'. If the specification were a very full one, then there would be relatively few people in each category – perhaps at most one, if the person's full name were specified, for example. On the other hand, a rather unspecific description would mean fewer categories, each with more members. The essential point is that we assume a level of specification has been fixed, by circumstance or design, and that people are assigned to different categories if and only if they can be distinguished on the basis of that specification. So the separation into categories ω_k represents the completest breakdown possible on the observations available.

It is useful to regard the categories $\omega_1, \omega_2, \ldots, \omega_K$ as points in an abstract space Ω, the *sample space*.

Now we introduce the idea of an *observable*, which is, briefly, any qualitative or quantitative property of an individual whose value is determined by the knowledge of the category to which he belongs. In the example above, 'marital status' and 'size of family' are observables. 'Year of birth' is an observable, but 'age' is not, because the exact instant of birth is not part of the specification. 'Income tax payable' is very nearly an observable, but not quite, because to be able to determine this we would need rather more information, such as ages of children, and reasons for special allowances.

Since an observable is something whose value is determined by the category, we can write it as a function $X(\omega)$, where ω takes the values $\omega_1, \omega_2, \ldots, \omega_K$. That is, *an observable is a function on the sample space*. For example, 'marital status' is the function which takes the values 'single', 'married', 'widowed', 'divorced', etc., appropriately, as ω takes the different category values ω_k.

Now, consider a numerically valued observable, such as 'size of family'. In summarizing the results of the census, one will often quote the average values of such variables; the average being the conventional arithmetic mean, with each individual equally weighted:

$$A(X) = \frac{1}{n} \sum_{k=1}^{K} n_k X(\omega_k) = \sum p_k X(\omega_k). \qquad \textbf{1.3.1}$$

Here we have denoted the number of people in the kth category by n_k, and the proportion n_k/n in that category by p_k. The notation $A(X)$ emphasizes the fact that the average is a figure whose value depends upon the particular observable X we are considering. In fact, $A(X)$ is a *functional* of the function $X(\omega)$, a quantity whose value is determined from the values of $X(\omega)$ by the rule **1.3.1**.

Although only numerical observables can be averaged, one can often convert non-numerical observables into numerical form. Consider for example, the observable 'marital status'. We could define the observable

$$X(\omega) = \begin{cases} 1 & \text{if the category } \omega \text{ is one of married people,} \\ 0 & \text{otherwise;} \end{cases} \qquad \textbf{1.3.2}$$

and $A(X)$ would then be the proportion of married people in the country.

The function defined in **1.3.2** is an *indicator function*, a function taking the value 1 in a certain ω-set (the 'married' set) and 0 elsewhere. Its average value is just the proportion of people falling into this set. This is the fact that we noted at the end of section 1.1: the proportion of the population which is married is the average value of the indicator function of the 'married' set.

We shall take the concept of an average as basic, so the properties of the functional $A(X)$ are important. The reader will easily deduce the following list of properties from formula **1.3.1**:

(i)   If $X \geqslant 0$, then $A(X) \geqslant 0$.

(ii)   If X_1 and X_2 are numerical observables, and c_1 and c_2 are constants, then

$$A(c_1 X_1 + c_2 X_2) = c_1 A(X_1) + c_2 A(X_2).$$

(iii) $A(1) = 1$.

In words, the operator A which determines the average $A(X)$ of an observable X is a positive linear operator, fulfilling the normalization condition (iii).

Instead of defining the averaging operator A explicitly by formula **1.3.1**, and then deducing properties (i–iii) from it, we could have gone the other way. That is, we could have regarded properties (i–iii) as those which we would intuitively expect of an averaging operator, and taken them as the axioms for a theory of such operators. Actually, in the present case, the approach taken scarcely makes much difference, because it follows from (i–iii) that the operator must have the form of rule **1.3.1** (see exercise 1.3.3). However, in more general cases there are great merits in the axiomatic approach, which is, indeed, the one upon which we base our treatment in section 2.2.

Exercises

1.3.1   All inhabitants of a town answer the following questionnaire:

   (a) Do you suffer from bronchitis?
   (b) If so, do you smoke?

How many points are there in the sample space? Could one decide on the basis of the answers to these questions whether a person was a smoker, and, if so, whether he was a bronchitic? (That is, are these properties observables?)

1.3.2   The set of values of an observable $X(\omega)$ on Ω form the points of a new sample space, Ω_X. Show that ω is observable on Ω_X (i.e. no information has been lost by the transformation) if and only if the K values $X(\omega_k)$ are distinct.

1.3.3   Suppose it is known that the sample space Ω consists of just K points $\omega_1, \omega_2, \ldots, \omega_K$, and that properties (i–iii) hold for the average of any observable X on this sample space. Show, by choosing $X(\omega)$ as the indicator function of an appropriate set, that $A(X)$ must have the form **1.3.1**, where the p_k are some set of numbers satisfying

$$p_k \geq 0 \quad (k = 1, 2, \ldots, K) \qquad \sum p_k = 1.$$

1.3.4   If c is a constant, show that

$$A[(X-c)^2] = A(X^2) - 2cA(X) + c^2.$$

Show, by minimizing this expression with respect to c, that

$$A(X^2) \geq [A(X)]^2.$$

What is the minimizing value?

1.4   Repeated experiments; expectation

Suppose that we carry out an experiment (or take an observation; we are using the word 'experiment' in a very general sense), and let all the possible distinguishable outcomes of the experiment be labelled by a variable ω. If the experiment has only K possible outcomes, then there will be just K possible ω-values, $\omega_1, \omega_2, \ldots, \omega_K$, as in the previous section. However, we must now be prepared to consider much more general situations (see section 1.5).

   Now, the experiment can be repeated: we might perform it n times, observing outcomes $\omega^{(1)}, \omega^{(2)}, \ldots, \omega^{(n)}$, where each $\omega^{(j)}$ corresponds to some point ω

in Ω, the space of possible outcomes. These n observations are rather like the n observations we made in taking a census of a country of n people, in the last section.

There is an important difference, however. In the case of the census, we had made a complete enumeration of the population, and could go no further (if interest is restricted to the population of that one country). In the case of the experiment, we could go on repeating observations indefinitely, and, provided the outcome of an experiment could not be completely predicted from those of previous experiments, each observation would be essentially new.

In a terminology and manner of thinking which have fallen out of fashion, but which are nevertheless useful, we imagine that we are sampling from a 'hypothetical infinite population', as compared with the physical finite population of section 1.3. But, just because the population is infinite, and complete enumeration is impossible, we cannot write down the 'total population average' as we did in formula **1.3.1**. However, the empirical fact that sample averages had to 'converge' leads us to postulate that the 'total population average' or 'long-term average' of an observable X does exist in principle. This idealized value we shall term the *expected value* or *expectation* of X, denoted $E(X)$, and we shall demand that the functional $E(X)$ have just the properties (i–iii) which we required of a finite population average $A(X)$ in section 1.3. Thus we have the basis of an axiomatic treatment, which is the modern approach to the subject.

We have appealed to the empirical fact that a sample average 'converges', i.e. that a partial average can approximate to the total average, if only the sample is large enough. This is useful, not only as an indication that the concept of total average is reasonable even when the average cannot be directly observed, but also as a practical procedure. For example, although complete enumeration is certainly possible if a census of a country is required, one will often carry out only a partial census (the census of a sample), on the grounds of economy. If only the sample is large enough, and precautions taken to make it representative, one can expect the results of the partial census to differ little from those of a complete census. (See exercises 4.3.3 and 4.5.5.)

1.5 More on sample spaces and observables

We retain the idea of the sample space Ω as a space of points ω, where ω indexes the possible outcomes of an experiment. We require that, if two possible outcomes are distinguishable, then the corresponding ω-values should also be distinct. Quotation of an ω-value thus specifies the outcome of the experiment completely, at the level of description adopted.

So, suppose we are observing the result of a football match. One level of description might be simply to report 'win, lose or draw' for the home team, so that Ω would contain just three points. A more refined level of description would be to give the score, so that ω would be a pair of integers (s_1, s_2), and Ω the set of such integer pairs.

This sample space is already more general than the one considered in section 1.3, in that it contains an infinite number of points. Of course, for practical purposes it is finite, since no team is ever going to make an infinite score. Nevertheless, there is no very obvious upper bound to the possible score, so it is best to retain the idea in principle that s_1 and s_2 can take any integral value.

For another example, suppose the experiment is one in which a recorder observes wind-direction in angular degrees. If we denote the result by θ, then θ is a number taking any value in $[0, 360)$. The natural sample space is thus a finite interval (or, even more naturally, the circumference of a circle). Again, Ω is infinite, in that it contains an infinite number of points. Of course, one can argue that, in practice, wind-direction can only be measured to a certain accuracy (say, to the nearest degree), so that the number of distinguishable experimental outcomes is in fact finite. However, it is actually simpler in many cases to allow that infinite accuracy of observation might be possible (see exercises 3 and 4), and to retain the continuous sample space of all numbers in $[0, 360)$.

A more refined experiment would be to observe both wind-direction θ and wind-speed v, so that Ω would be the set of values

$$\{\omega = (\theta, v) \quad 0 \leqslant \theta < 360, v \geqslant 0\}:$$

a two-dimensional infinite sample space.

As in section 3, we define an *observable* as any quantity whose value (not necessarily numerical) is determined once the outcome of the experiment is known. It can thus again be regarded as a function $X(\omega)$ on Ω.

So, in the case of the football match, the 'winning margin' $|s_1 - s_2|$ is observable on the second space, but not on the first. In the case of the wind-measurement, the 'component of wind-velocity in direction α', $v \cos\{\pi(\alpha - \theta)/180\}$, is observable on the second space, but not on the first.

Exercises

1.5.1   Suppose the 'experiment' is the taking of a vote on a motion by a committee of m members. Each person can vote for or against the motion, or can abstain. What is the number of points in Ω if the voting is (i) open, (ii) by secret ballot? A second vote on the same issue would be a repetition of the experiment. This would be unpredictable from the first, and so a genuinely new observation, only if people are prepared to change their minds on the issue. Supposing they are, how many points are there in the sample space for the composite experiment with two votes taken?

1.5.2   Note that if one is interested in discussing simultaneously all n outcomes of an experiment repeated n times, then one essentially has a compound observation, and must construct a compound sample space. The tossing of a coin needs a sample space of 2 points; if one discusses the results of n tosses one needs a sample space of 2^n points. Is the variable 'number of tosses before the first head is thrown' observable on this latter sample space?

1.5.3 Note that, if we allow explicitly for experimental error in the construction of Ω, the description of Ω becomes heavily dependent on the precise method of experiment. For example, suppose we measure wind-velocity (i) by measuring components of velocity along two given orthogonal axes to the nearest mile per hour, or (ii) by measuring direction to nearest degree and speed to the nearest mile per hour. Then the two sample spaces are different. In general, we encounter trouble then with variable transformations, and 'point densities' which vary over Ω. Yet it may be that in a fundamental study it would be the method and accuracy of observation which determined the structure of Ω.

1.5.4 Suppose that one tries to observe a quantity y, but actually observes $y' = y + \varepsilon$, where ε is an observational error, of which one knows only that $|\varepsilon| < d$. The variable 'y to within $\pm d$' is thus observable by definition. Show, however, that 'y to the nearest multiple of δ' is not observable for any δ.

1.5.5 Suppose one observed temperature θ as a function $\theta(t)$ of time over a period of twenty-four hours. Note that the sample space for this experiment must have a continuous infinity of dimensions, at least if one postulates no continuity properties for $\theta(t)$.

1.5.6 An electrical circuit consists of m elements in series, and exactly one element is known to be faulty. A test between any two points in the circuit shows whether the fault lies between the two points or not. Let r be the minimum number of tests needed to locate the fault, i.e. to make the variable 'location of the fault' observable. What is r if $m = 8$? Show that for large m,

$$r \simeq \log_2 m$$

Note that the r tests are not to be regarded as the r-fold repetition of an experiment, but rather as a single compound experiment.

2 Expectation

2.1 Random variables

We shall now take up in earnest the axiomatic treatment sketched in section 1.3. Let us postulate that with each numerically valued observable $X(\omega)$ can be associated a number $E(X)$, the *expected value* or *expectation* of X. A situation in which this is possible will be called a *probability process*, and $X(\omega)$ will be termed a *random variable* (henceforth abbreviated to r.v.).

We are in effect changing our viewpoint now, from considering what has happened in an experiment already performed, to considering what migh. happen in an experiment yet to be performed. The expectation $E(X)$ is the idealized average of X, the average being taken over all the outcomes that might result if the experiment were actually performed. The idea of such an average is given empirical support by the fact that sample averages seem to 'converge' to a limit with increasing sample size. This is of course, only an empirical fact, which motivates the concept of an idealized average, but does not justify it. However, the achievement of a self-consistent and seemingly realistic theory is justification enough in itself. As an example of 'realism', we shall find that convergence of sample averages to expectation values is a feature which is reproduced in certain quite general models, under plausible assumptions concerning the joint outcomes of repeated experiments (sections 6.1, 6.2, 9.5).

We shall postulate in section 2.2 that the expectation operator E has certain axiomatic properties, essentially the properties (i–iii) asserted for the averaging operator A in section 1.3. These axioms, although few and simple, will take us quite a long way.

In particular, they will enable us to develop the consequences of the basic assumptions of a physical model, and this is the usual aim. For example: if one makes some probabilistic assumptions concerning the fission of uranium nuclei by incident neutrons, what can one then infer concerning the behaviour of an atomic reactor core? If one makes some probabilistic assumptions concerning the development of the economy from day to day, what can one infer concerning the state of the economy three months hence?

From an abstract point of view the situation is as follows. By physical or other arguments one is given the expectation values of a family $\mathscr{F}$ of r.v.s X_ν. From this information one would wish to determine as closely as possible the expectation values of other r.v.s of interest, Y_ν, by appeal to the axioms.

Can an expectation be defined for any r.v.? We shall take the view that it can, in so far as that, once we have established that the given expectations $E(X_v)$ are internally consistent with the axioms, we shall thereafter accept any statement concerning an expectation $E(Y)$ which can be derived via the axioms from the given expectations, for any r.v. Y. Such an approach can be faulted only if such derived statements turn out to contain an inconsistency, and this we shall not find. It does not imply that $E(X)$ can be simultaneously and consistently prescribed for all X on Ω; cases are in fact known for which this is impossible.

Of course, if Y is, relative to the X_v, rather bizarre as a function of ω, then the bounds on $E(Y)$ derivable from the $E(X_v)$ will presumably be rather wide, and $E(Y)$ may even be completely indeterminate.

The delineation of the class $\mathscr{F}_1$ of r.v.s whose expectations are exactly determined by the given values is the *extension* problem, considered in chapter 10. Sometimes the term *random variable* is reserved for members of $\mathscr{F}_1$, or used in an even narrower sense (section 3.5). This is not an essential point: to use the term in the present broader sense is convenient, and should not cause one to forget that the class of r.v.s whose expectations are actually known may be relatively small.

Note that the values $\pm\infty$ for an expectation have not been excluded as in any sense improper.

2.2 Axioms for the expectation operator

We shall refer indiscriminately to E as the expectation operator, or to $E(X)$ as the expectation functional. The point is, that E determines a number $E(X)$ from a function $X(\omega)$.

The exact form of the operator, i.e. the actual rule for attaching an expectation $E(X)$ to a r.v. X, must be determined by special arguments in individual cases and these arguments will usually determine $E(X)$ only for certain X. Much of the rest of the book will be concerned with such particular cases: physical processes of one sort or another. However, in this section we shall concern ourselves with the general rules which E should obey if it is to correspond to one's intuitive idea of an averaging operator. These rules will take the form of axioms, relating the expectations of r.v.s.

Axiom 1   *If $X \geqslant 0$ then $E(X) \geqslant 0$.*

Axiom 2 $E(cX) = cE(X)$ *for c a constant.*

Axiom 3 $E(X_1 + X_2) = E(X_1) + E(X_2)$.

Axiom 4 $E(1) = 1$.

Axiom 5   *If a sequence of r.v.s $\{X_n(\omega)\}$ increases monotonically to a limit $X(\omega)$ then $E(X) = \lim E(X_n)$.*

The first four axioms state that E is a positive, linear operator with the normalization $E(1) = 1$, just as was the averaging operator (equation **1.3.1**). Axiom 5 is a continuity demand, stating that for a monotone sequence of r.v.s the operations E and lim commute. Although this condition is also satisfied by the averaging operator defined in **1.3.1**, as an axiom it appears somewhat less natural than the others, particularly since a weak form of it can be derived from Axioms 1–4 (see exercise 2.2.5). In fact, one can go a long way without it, and there are interesting physical situations for which the axiom does not hold (see exercise 2.4.3). However, some condition of this type becomes necessary when one considers limits of infinite sequences, as we shall have occasion to do later (sections 9.7 and 10.3).

The axioms have certain immediate consequences that the reader can verify; for example, that

$$E\left(\sum_1^n c_j X_j\right) = \sum_1^n c_j E(X_j),$$ **2.2.1**

if the c_js are constants and n is finite, and that

$$X_1 \leqslant Y \leqslant X_2$$ **2.2.2**

implies

$$E(X_1) \leqslant E(Y) \leqslant E(X_2).$$ **2.2.3**

The equations in the axioms are all to be understood in the sense that, if the right-hand member is well-defined, then the left-hand member is also well-defined, and the two are equal. There are occasional failures. For example, suppose that $E(X_1) = +\infty$ and $E(X_2) = -\infty$. Then Axiom 3 would give $E(X_1 + X_2)$ the indeterminate value $+\infty - \infty$.

We can avoid such indeterminacies by restricting the class of r.v.s considered. For example, suppose we separate X into positive and negative parts:

$$X = X_+ - X_-$$

where

$$X_+ = \begin{cases} X & (X \geqslant 0), \\ 0 & (\text{otherwise}), \end{cases}$$ **2.2.4**

and require that $E(X_+) < \infty$, $E(X_-) < \infty$. Since $|X| = X_+ + X_-$, this is equivalent to requiring that

$$E(|X|) < \infty.$$ **2.2.5**

If we restrict ourselves to r.v.s for which inequality **2.2.5** is true (r.v.s with *finite absolute expectation*) then $\sum_1^n E(X_j)$ will always be well-defined. This is a convenient restriction, on the whole, and we shall henceforth adopt it, unless the contrary is stated.

Exercises

2.2.1 Show that Axioms 1 and 4 are equivalent to the single statement

$$a \leqslant X \leqslant b \quad \text{implies} \quad a \leqslant E(X) \leqslant b,$$

for a, b constant. Although such a single axiom would be more economic, it is perhaps helpful to separate the properties of positivity and normalization.

2.2.2 Show that equation **2.2.1** holds for n infinite, if all the $c_j X_j$ are of the same sign.

2.2.3 Show that $|E(X)| \leqslant E(|X|)$.

2.2.4 Show that $E(|X_1 + X_2|) \leqslant E(|X_1|) + E(|X_2|)$.

2.2.5 Show, without appeal to Axiom 5, that if

$$|X_n - X| \leqslant Y_n \quad \text{and} \quad E(Y_n) \to 0,$$

then $E(X_n) \to E(X)$.

2.3 Events; probability

By an *event* A we shall understand a set of possible experimental outcomes. This will correspond to a set of points in Ω, which will also be denoted by A.

Thus, for the football example of section 1.5 we might consider the event 'the home team won'. In the first sample space suggested there this would correspond to a single point; in the second sample space it would correspond to the set of points satisfying $s_1 > s_2$.

For the wind-velocity example, we might consider the event 'the wind-speed exceeds 50 m.p.h.'. This would be observable only on the second sample space, when it would correspond to the set of points $v > 50$.

The *probability of* A, denoted $P(A)$, will be defined as

$$P(A) = E(I_A), \tag{2.3.1}$$

where $I_A(\omega)$ is the *indicator function* of the set A:

$$I_A(\omega) = \begin{cases} 1 & (\omega \in A), \\ 0 & (\omega \notin A). \end{cases} \tag{2.3.2}$$

$P(A)$ is to be regarded as the expected proportion of experiments in which A actually occurs. The motivation for the definition comes from the finite population census of section 1.3, where we saw that the proportion of the population falling in a set A was just the average of the indicator variable for the set A.

We shall not investigate the concepts of event or probability to any extent before chapter 3, but it is helpful to have them formulated.

The probability measure $P(A)$ is a function with a *set* A as argument. However, it is notationally very convenient to take sometimes just a description of the event as the argument, regarding this as equivalent to the corresponding set. Thus we write the probability that X is greater than Y as $P(X > Y)$, rather than the more correct but cumbersome $P(\{\omega : X(\omega) > Y(\omega)\})$. The same goes for more verbal descriptions: we would write $P(\text{rain})$ rather than $P(\text{the set of } \omega$ for which it rains). Nevertheless, the true argument of $P(\quad)$ is always a set in Ω.

Exercise

2.3.1   Show that $0 \leqslant P(A) \leqslant 1$, and interpret the extreme cases.

2.4   Some examples of an expectation

Before proceeding further, we should show that expectation operators satisfying the axioms of section 2.2 really do exist. It is sufficient to find some examples, and for these we can choose the types of process arising in applications.

Consider first the analogue of **1.3.1**, and suppose that $E(X)$ is given by the formula

$$E(X) = \sum_k p_k X(\omega_k) \qquad\qquad\qquad \textbf{2.4.1}$$

for any r.v. $X(\omega)$ for which the sum is absolutely convergent. Then E will satisfy all the axioms if the p_k are non-negative and if $\sum p_k = 1$, as the reader can verify.

These conditions are also necessary. For, setting $X(\omega) = I_A(\omega)$ in **2.4.1**, we find that

$$P(A) = \sum_a p_k, \qquad\qquad\qquad \textbf{2.4.2}$$

where a is the set of values of k for which ω_k belongs to A. Expression **2.4.2** must be non-negative, since it is the expectation of a non-negative r.v. Letting A consist of the single point ω_k, we find

$$p_k = P(\omega_k),$$

so that p_k is non-negative, and is identified as the probability of the experimental outcome associated with ω_k. Letting $A = \Omega$, we find $\sum p_k = E(1) = 1$.

If A is a set which contains none of the values ω_k at all, we see from **2.4.2** that $P(A) = 0$. We can thus characterize the process corresponding to **2.4.1** as one in which the only possible experimental outcomes are those corresponding to the points ω_k, these having respective probabilities p_k. One would say that there is a *discrete probability distribution* over Ω, concentrated on the points ω_k.

As an illustration, suppose one throws a die and observes the number occurring; this number can itself be denoted by ω. Let Ω be the real line, that is, let us allow the possibility that ω could take any real value. In fact, however, the only values possible are $\omega = 1, 2, 3, 4, 5, 6$. If we assume that the die is fair, then by symmetry all the p_k should be equal, and consequently equal to $\frac{1}{6}$. Hence in this case

$$E(X) = \tfrac{1}{6} \sum_{k=1}^{6} X(k).$$

The fact that in this formula $X(\omega)$ has no argument other than the values $1, 2, \ldots, 6$ indicates that these are the only values possible: the fact that all the p_k are equal expresses the symmetry of the die.

For a second type of process, let us suppose, for convenience rather than necessity, that the sample space is the real line, so that ω is a real number. Suppose that

$$E(X) = \int_{-\infty}^{\infty} X(\omega) f(\omega)\, d\omega \qquad\qquad \textbf{2.4.3}$$

for all r.v.s of the class $\mathscr{D}$ for which the integral is defined and absolutely convergent. Then E will obey the axioms (at least for r.v.s of $\mathscr{D}$; see the note at the end of the section) if f obeys the conditions

$$f(\omega) \geqslant 0, \qquad\qquad \textbf{2.4.4}$$

$$\int_{\Omega} f(\omega)\, d\omega = 1. \qquad\qquad \textbf{2.4.5}$$

The relation analogous to expression **2.4.2** is

$$P(A) = E(I_A) = \int_A f(\omega)\, d\omega, \qquad\qquad \textbf{2.4.6}$$

so that $f(\omega)$ can be regarded as a *probability density* on Ω. In this case one speaks of a *continuous probability distribution* on Ω. The idea can be extended to more general sample spaces than the real line, provided one has an appropriate definition of the integral **2.4.3**.

As an example, consider the spinning of a roulette wheel. If ω is the angle in radians that the pointer makes with some reference radius on the wheel when the wheel comes to rest, then ω can take values only in the range $[0, 2\pi)$. If the wheel is a fair one, all these values will be equally likely, so, by symmetry, the expectation formula must be

$$E(X) = \frac{1}{2\pi} \int_{0}^{2\pi} X(\omega)\, d\omega.$$

That is, we have a continuous probability distribution with

$$f(\omega) = \begin{cases} \dfrac{1}{2\pi} & (0 \leqslant \omega < 2\pi), \\ 0 & (\text{otherwise}). \end{cases} \qquad \textbf{2.4.7}$$

This example can help us to clarify a point: the distinction between impossible events, and events of zero probability. Impossible events (for example, the throwing of a seven with a die) have zero probability: the converse is not necessarily true. For, consider the probability that the roulette wheel comes to rest within an angle δ of a prescribed direction θ: the probability of this is

$$P(\theta - \delta < \omega < \theta + \delta) = \frac{\delta}{\pi},$$

for $\delta < \pi$. As δ tends to zero this probability also tends to zero. In other words the event $\omega = \theta$, that the rest-angle ω has a prescribed value, has zero probability. Yet this event is plainly not impossible. The event has zero probability, not because it is impossible, but because it is just one of an infinite number of equally probable experimental outcomes.

If a particular r.v. $X(\omega)$ is such that

$$E[H(X)] = \int H(x)f(x)\,dx \qquad \textbf{2.4.8}$$

for any function H for which the integral is meaningful and convergent, then we have a distribution which is continuous on the sample space Ω_x constituted by the x-axis. In this case the r.v. $X(\omega)$ is said to be continuously distributed with *probability density function* (or *frequency function*) $f(x)$. This is very much the same situation as before: we have, for example

$$P(X \in A) = \int_A f(x)\,dx, \qquad \textbf{2.4.9}$$

except that the continuous distribution is a property of the r.v. X rather than of the basic sample space Ω. It is conventional to use an upper case letter X for the r.v., and a corresponding lower case letter x for particular values the r.v. may adopt, and on the whole this is a helpful distinction. Thus, we write $f(x)$ rather than $f(X)$.

Note. Representation **2.4.3** is restricted to a class of r.v.s $\mathscr{D}$ because the integral is presumably to be interpreted in some classical sense, such as the Riemann sense, and Xf must then be Riemann integrable. However, use of the axioms will enable one to construct bounds for expectations of r.v.s X for which Xf is not Riemann integrable, even possibly to the point of determining $E(X)$ completely. So, $E(X)$ is not necessarily representable as a Riemann integral for all expectations which can be derived from representation **2.4.3**.

Exercises

2.4.1   Suppose that a person joining a queue has to wait for a time $X(\omega)$ before he is served, and that

$$E[H(X)] = pH(0) + \int_0^\infty H(x)f(x)\,dx$$

for all functions H for which this expression is defined and convergent. Find the conditions on p and f for this formula to represent an expectation on Ω_X and interpret the formula.

2.4.2   An electron oscillating in a force field has energy ε which can take values

$$\varepsilon_k = \alpha(k + \tfrac{1}{2})$$

with probabilities proportional to

$$\exp(-\beta\varepsilon_k) \quad (k = 0, 1, 2, \dots),$$

where α and β are constants. Determine $E(\varepsilon)$ and $E(\varepsilon^2)$.

2.4.3   Suppose that the expectation operator is defined by

$$E(X) = \lim_{D \to \infty} \frac{1}{2D} \int_{-D}^{D} X(\omega)\,d\omega.$$

Show that this satisfies the first four axioms, but not the fifth. (Consider the sequence of r.v.s

$$X_n(\omega) = \begin{cases} 1 & (|\omega| \leqslant n), \\ 0 & (|\omega| > n), \end{cases}$$

for $n = 0, 1, 2 \dots$.) This process would correspond to a uniform distribution over the whole infinite axis, and so might, for example, be used to represent the position of a star equally likely to lie anywhere within an infinite universe.

2.4.4   Show that if $P(X \leqslant x)$ is differentiable in x, then X has a probability density, which is equal to

$$f(x) = \frac{\partial}{\partial x}P(X \leqslant x).$$

This is the relation inverse to relation **2.4.9**.

2.4.5   Let ω_1 and ω_2 be the rest-angles observed in two consecutive spins of a roulette wheel, and suppose that expectations on the two-dimensional sample space thus generated are given by

$$E(X) = \frac{1}{4\pi^2} \int_0^{2\pi} \int_0^{2\pi} X(\omega_1, \omega_2)\,d\omega_1\,d\omega_2.$$

Interpret this formula. If $X = \omega_1 + \omega_2$, then show, either from exercise 2.4.4 or by calculating $E[H(X)]$ for arbitrary H, that X is continuously distributed with density

$$f(x) = \begin{cases} \dfrac{x}{4\pi^2} & (0 \leqslant x \leqslant 2\pi), \\[2mm] \dfrac{4\pi - x}{4\pi^2} & (2\pi \leqslant x < 4\pi), \\[2mm] 0 & \text{(otherwise).} \end{cases}$$

2.5 Applications: optimization problems

The theory we have developed, slight though it is as yet, is enough to help us to useful conclusions in a variety of problems. Of particular interest are those problems concerned with *optimization*, where one is trying to achieve maximum expected return in some enterprise.

A typical optimization problem is that of the newsagent who stocks N copies of a daily paper, and wishes to choose N so as to maximize his expected profit. Let a be the profit on a paper which is sold, b the loss on an unsold paper, and c the loss if a customer wishes to buy a paper when stocks are exhausted. The quantities a and b can be determined immediately from the wholesale and retail prices of the newspaper. The quantity c is less easy to determine, because it measures 'loss of goodwill due to one lost sale' in monetary terms, but an estimate of it must be made if the situation is to be analysed.

If the newsagent stocks N papers and has X customers on a given day, then the components of his profit (or negative loss) are as follows:

	Profit	
Item	$X \leqslant N$	$X > N$
Sales	aX	aN
Unsold papers	$-b(N - X)$	0
Unsatisfied demand	0	$-c(X - N)$

Thus his net profit is

$$g_N(X) = \begin{cases} (a+b)X - bN & (X \leqslant N), \\ (a+c)N - cX & (X > N). \end{cases}$$

If X were known, then he would obviously maximize profit by choosing $N = X$. However, the demand X will certainly be variable. and can realistically be regarded as a r.v., so that one has instead to work with *expected* profit

$$G_N = E[g_N(X)]$$

and choose N so as to maximize this. If expected profit is virtually identical with long-term average profit (the intuitive foundation of our theory; see also sections 6.1 and 6.2), then to maximize G_N is a reasonable procedure.

The change in expected profit when an extra paper is stocked is

$$G_{N+1} - G_N = E[g_{N+1}(X) - g_N(X)]$$
$$= E[-b + (a+b+c)\mathscr{H}(X-N)]$$

where $\mathscr{H}$ is the discontinuous Heaviside function

$$\mathscr{H}(X) = \begin{cases} 1 & (X > 0), \\ 0 & (X \leqslant 0). \end{cases} \qquad \textbf{2.5.1}$$

The increment in expected profit is thus

$$G_{N+1} - G_N = -b + (a+b+c)E[\mathscr{H}(X-N)]$$
$$= -b + (a+b+c)P(X > N),$$

since $\mathscr{H}(X-N)$ is the indicator function of the set $X > N$. For small enough N this quantity is positive, but ultimately it turns negative, and the first value of N for which it does so is the optimal one. Roughly, one can say that the optimal N is the root of the equation $G_N \simeq G_{N+1}$, or

$$P(X > N) \simeq \frac{b}{a+b+c}.$$

To complete the solution of the problem one needs to know $P(X > N)$ as a function of N. In practice, one would use records of past sales to obtain an estimate of this function. For example, $P(X > N)$ could be estimated directly from the actual proportion of times over a long period that potential sales have exceeded N. More refined methods are possible if one can restrict the form of the function $P(X > N)$ on theoretical grounds.

The treatment of this problem has not followed quite the course we promised in section 2.1, in that we have simply ploughed ahead with maximization of expected profit, assuming that all expectations required for this purpose, such as $P(X > N)$, were known. For cases where one has to make the best of less information, see exercise 10.1.3.

This example was simplified by the fact that a newspaper can be assumed to have commercial value only on the day of issue, so that there is no point in carrying stock on from one day to another; each day begins afresh. However, suppose that the newsagent also stocks cigarettes. For this commodity he will certainly carry stock over from day to day, and the decisions made on a given day will be influenced by those made previously. So, in taking a decision, he must think of its implications for the future, as well as for the present. This much more difficult and interesting situation, in which one makes a sequence of interacting decisions, optimally if practicable, has given rise to the technique of *dynamic programming*. We return to this topic in section 12.3.

Exercises

2.5.1 Suppose that one incurs a loss as if early for an appointment by time s, and a loss bs if late by time s $(s \geqslant 0)$. The time taken to reach the place of appointment is a continuously distributed r.v., T. Suppose that one leaves for this place a time t before the appointed instant. Show that the value of t minimizing expected loss is determined by

$$a P(T < t) = b P(T > t) \quad \text{or} \quad P(T < t) = \frac{b}{a+b}.$$

2.5.2 A steel billet is trimmed to length x and then rolled. After rolling, its length becomes $y = \beta x + \varepsilon$, where ε is a r.v. expressing the variability of rolling. It is then trimmed again to a final length z. If y is greater than z there is a loss proportional to the excess: $a(y-z)$. If y is less than z then the billet must be remelted, and there is a flat loss b. The original trim-length x must now be chosen so as to minimize expected loss. Show that if ε has probability density $f(\varepsilon)$ then the equation determining the optimal value of x is

$$bf(z - \beta x) = a \int_{z - \beta x}^{\infty} f(\varepsilon)\, d\varepsilon.$$

2.6 Applications: least square approximation of random variables

Let us consider an optimization problem of rather a different character. Commonly one is in the position of being able to observe the values of a number of r.v.s, $X_1, X_2, \ldots, X_m$, and wishes to infer from these the probable value of another r.v., Y, which cannot itself be observed. For example, one may wish to predict the position Y of an aircraft, at some future time, on the basis of observations X_j already made upon its path. This is a prediction problem, such as also occurs in numerical weather forecasting, and medical prognosis. The element of prediction in time need not always be present, so it is perhaps better to speak of 'approximation'. For instance, the r.v.s X_j might represent surface measurements made by an oil prospector, and Y a variable related to the presence of oil at some depth under the earth's surface.

In any case, one wishes to find a function of the X_js

$$\hat{Y} = \phi(X_1, X_2, \ldots, X_m)$$

which approximates Y as well as possible. That is, ϕ is to be chosen so that $\hat{Y}$ lies as near in value to Y as possible, in some sense.

An approach very common in both theoretical and practical work is to restrict oneself to *linear approximants*

$$\hat{Y} = \sum_{k=1}^{m} a_k X_k, \qquad\qquad\qquad \textbf{2.6.1}$$

and to choose the coefficients a_k so as to minimize the *mean square error*

$$\Delta = E[(Y-\hat{Y})^2] = E\left[\left(Y-\sum_1^m a_k X_k\right)^2\right]$$

$$= E(Y^2) - 2\sum_k a_k E(X_k Y) + \sum_j \sum_k a_j a_k E(X_j X_k). \qquad \textbf{2.6.2}$$

The r.v. $\hat{Y}$ is then known as the *linear least square approximant* (or predictor) of Y in terms of $X_1, X_2, \ldots, X_m$. This is a valuable technique practically, one reason for this being, as we see from equation **2.6.2**, that in order to apply it we need only know the values of the expectations of the squares and products of the r.v.s concerned. In practice, these expectations would be estimated from long-period averages of the corresponding quantities over past data.

We can rewrite **2.6.2** in matrix form as

$$\Delta = U_{YY} - 2a'U_{XY} + a'U_{XX}a$$

say, where we have defined a scalar U_{YY}, column vectors a and U_{XY}, and a matrix U_{XX}. By equating the derivatives $\partial\Delta/\partial a_j$ to zero we obtain the equations determining the optimal set of coefficients:

$$U_{XX}a = U_{XY}. \qquad \textbf{2.6.3}$$

The stationarity condition, equation **2.6.3**, does in fact locate a minimum of Δ with respect to a: see exercise 2.6.2.

If the matrix U_{XX} is non-singular, then we can solve the system **2.6.3** to obtain

$$a = U_{XX}^{-1} U_{XY}. \qquad \textbf{2.6.4}$$

An approximant which incorporates a constant term,

$$\hat{Y} = a_0 + \sum_1^m a_k X_k, \qquad \textbf{2.6.5}$$

seems to be more general than expression **2.6.1**, but in fact is not. The right-hand member of expression **2.6.5** can be written $\sum_0^m a_k X_k$, where X_0 is a r.v. identically equal to 1, so that **2.6.5** is in fact of the same form as **2.6.1**.

The extra minimization equation (w.r.t. a_0) is

$$E(Y) = a_0 + \sum_1^m a_k E(X_k),$$

so that we can rewrite expression **2.6.5** in the form

$$\hat{Y} - E(Y) = \sum_1^m a_k[X_k - E(X_k)]. \qquad \textbf{2.6.6}$$

As far as the determination of $a_1, a_2, \ldots, a_m$ is concerned, the whole previous analysis can be repeated, with Y and X_k replaced by $Y - E(Y)$ and $X_k - E(X_k)$

respectively (see exercise 2.6.4). Least square approximation is an important and recurrent topic: see sections 5.3, 9.7 and 10.4.

Exercises

2.6.1 Show that $c'U_{XX}c \geqslant 0$ for any real vector c.

2.6.2 Write Δ as $\Delta(a)$, to emphasize its dependence on the coefficient vector a. If $\hat{a}$ is any solution of system **2.6.3**, show that

$$\Delta(a) = \Delta(\hat{a}) + (a - \hat{a})'U_{XX}(a - \hat{a}),$$

so that any solution of **2.6.3** truly minimizes Δ.

2.6.3 Show that the minimized value of Δ can be written in the various forms

$$\Delta = U_{YY} - a'U_{XY} = U_{YY} - U_{YX}U_{XX}^{-1}U_{XY}$$

$$= \frac{\begin{vmatrix} U_{YY} & U_{YX} \\ U_{XY} & U_{XX} \end{vmatrix}}{|U_{XX}|}.$$

Here $U_{YX} = U'_{XY}$, and we have supposed U_{XX} non-singular.

2.6.4 Show that, for the approximating formula **2.6.5**, formulae **2.6.2**–4 hold, and also the formulae of exercise 2.6.3, if the *product-means* $E(X_jX_k)$ are replaced by the *covariances*

$$\text{cov}(X_j, X_k) = E[\{X_j - E(X_j)\}\{X_k - E(X_k)\}]$$

$$= E(X_jX_k) - E(X_j)E(X_k), \qquad \textbf{2.6.7}$$

and similarly for $E(Y^2)$, $E(X_jY)$. The covariance of two r.v.s provides a measure of the extent to which the deviations of the r.v.s from their respective mean values tend to vary together. Show that the equations determining the a_j can be written

$$\text{cov}(Y - \hat{Y}, X_j) = 0 \quad (j = 1, 2, \ldots, m).$$

2.6.5 Suppose that the midday temperature on day j is X_j ($j = \ldots, -2, -1, 0, 1, 2, \ldots$), and that records show that to a reasonable approximation

$$E(X_j) = \mu,$$

$$\text{cov}(X_j, X_k) = \alpha\beta^{|j-k|},$$

where μ, α and β are constants ($|\beta| < 1$).

Verify, by appeal to exercise 2.6.2, that

$$\hat{X}_{n+s} = \mu + \beta^s(X_n - \mu) \quad (s \geqslant 0)$$

is a linear least square predictor (with a constant term) of X_{n+s} based on

$X_n, X_{n-1}, X_{n-2}, \ldots$. This is an example of prediction s steps ahead in a *time series*.

2.7 Some implications of the axioms

As mentioned in section 2.1, it is typical that in the enunciation of a problem one is given the expectation values for some class $\mathscr{F}$ of r.v.s X, this class being in some cases quite small, in others larger. The axioms then appear as *consistency conditions* among the given expectations. For r.v.s outside $\mathscr{F}$ they still appear as consistency conditions, which restrict the possible values of the unknown expectations, sometimes, indeed, to the point of determining them completely. For example, if $\mathscr{F} = \{X_1, X_2, \ldots\}$ then the value of $E(Y)$ for $Y = \sum_1^n c_j X_j$ is determined uniquely by equation **2.2.1**, at least if we assume all $c_j X_j$ to have finite absolute expectation. The question of limits of such sums as $n \to \infty$ is more delicate, and will be treated in chapter 10.

Again, if $X_1 \leqslant Y \leqslant X_2$, then we have the implication **2.2.3**. If in particular $E(X_1) = E(X_2)$, then the value of $E(Y)$ is fully determined. (This does not imply $X_1(\omega) \equiv Y(\omega) \equiv X_2(\omega)$, when the result would be trivial; in case **2.4.1**, for example, the functions might differ for values of ω other than $\omega_1, \omega_2, \ldots$.)

The consistency conditions **2.2.1** and **2.2.3** have consequences which are not immediately obvious: in this section we shall follow up some of these. In chapter 10 we shall go even further, by considering limit results, and appealing to the further consistency condition implied by Axiom 5.

For instance, consider the obvious inequality

$$\mathscr{H}(X - a) \leqslant \frac{X}{a} \qquad (X \geqslant 0), \qquad\qquad \textbf{2.7.1}$$

where a is a positive constant, and $\mathscr{H}$ the Heaviside function defined in equation **2.5.1**. Suppose now that X is a non-negative r.v. Taking expectations on both sides of relation **2.7.1**, (as inequality **2.2.3** indicates we may) we obtain then

$$P(X > a) \leqslant \frac{E(X)}{a}. \qquad\qquad \textbf{2.7.2}$$

This simple result, known as the *Markov inequality*, is extremely useful. It implies that one can set an upper bound on the probability that X exceeds a given value if X is known to be positive and have finite expectation. The smaller $E(X)$, the smaller this bound, as one might expect.

If $E(X) \leqslant a$, then this inequality is also sharp, in the sense that one can find a process for which $E(X)$ has the prescribed value, and equality is attained in relation **2.7.2**. The process is that in which X takes only the two values 0 and $a+$ with probabilities $1 - E(X)/a$ and $E(X)/a$ respectively.

Consider now a r.v. X which is not restricted in sign, and for which the values of $E(X)$ and $E(X^2)$ are assumed known. By taking expectations in the inequality

$$\mathscr{H}(|X-b|-a) \leqslant \left(\frac{X-b}{a}\right)^2$$

we obtain the relation

$$P(|X-b| > a) \leqslant \frac{E[(X-b)^2]}{a^2}. \qquad \textbf{2.7.3}$$

That is, we have an upper bound in terms of $E(X)$ and $E(X^2)$ for the probability that X deviates from a value b by more than a given amount a.

The mean square

$$E[(X-b)^2] = E(X^2) - 2bE(X) + b^2$$

reaches its minimum value with respect to b when $b = E(X)$. Let us introduce a notation for the two important quantities that arise in this way:

$$\mu = E(X), \qquad \textbf{2.7.4}$$

$$\text{var}(X) = E[(X-\mu)^2] = E(X^2) - [E(X)]^2. \qquad \textbf{2.7.5}$$

The mean value of X, μ, can be regarded as a 'central value' of the X distribution. Quantity **2.7.5**, known as the *variance* of X, obviously gives a measure of the degree of spread of X about this central value. Setting $b = \mu$ in inequality **2.7.3** we obtain *Chebichev's inequality*:

$$P(|X-\mu| > a) \leqslant \frac{\text{var}(X)}{a^2}. \qquad \textbf{2.7.6}$$

This is again a most useful inequality, relating the probability of deviations fro, the mean value to the variance of X.

I' X has zero variance, so that $E[(X-\mu)^2] = 0$, we shall say that X *is equal to μ in mean-square*, written

$$X \overset{\text{m.s.}}{=} \mu. \qquad \textbf{2.7.7}$$

This has the implication that

$$E[H(X)] = H(\mu)$$

for any $H(X)$ bounded each way by a quadratic function in X taking the value $H(\mu)$ at $X = \mu$. (More specifically, the statement that $E[(X-\mu)^2]$ is arbitrarily small implies that $|E[H(X)] - H(\mu)|$ is arbitrarily small for such a function, although this is not necessarily true for other functions; see exercise 2.7.4.) In particular, as we see from inequality **2.7.6**, $P(|X-\mu| \leqslant a) = 1$ for arbitrarily small positive a, or $X = \mu$ *with probability one*.

Exercises

2.7.1   Show that if $M(X)$ is an increasing positive function of $|X|$, then

$$P(|X| > a) \leqslant \frac{E[M(X)]}{M(a)}.$$

2.7.2   Show that if $\text{var}(X) \leqslant a^2$ then Chebichev's inequality **2.7.6** is sharp, in the same sense as inequality **2.7.2** is. (Consider a process in which X can only assume the values $\mu, \mu \pm a +$.)

2.7.3   Show that inequality **2.7.2** fails if X may also take negative values.

2.7.4   Consider a r.v. X taking the values $0, n^{\frac{1}{4}}, -n^{\frac{1}{4}}$ with respective probabilities $1 - 1/n, 1/2n, 1/2n$. Show from this example that one may have $E(X^4) = 1$, although $\text{var}(X)$ is indefinitely small. Show that $E(X^4)$ may indeed adopt any value from zero to plus infinity, consistently with $\text{var}(X)$ being arbitrarily small.

Inequalities **2.7.2** and **2.7.3** are direct consequences of inequality **2.2.3**: the following important example leads to relations which are less evident. Suppose that for a set of r.v.s $X_1, X_2, \ldots, X_m$ we know the *product moments*

$$u_{jk} = E(X_j X_k)$$

already encountered in the least square approximation example of section 2.6. For any set of real constants c_j we have

$$0 \leqslant E[(\sum c_j X_j)^2] = \sum\sum c_j c_k u_{jk},$$

or $c'Uc \geqslant 0,$ **2.7.8**

if U is the matrix with elements u_{jk}, and c the vector with elements c_j.

The *product moment matrix* U is evidently symmetric, and we see from inequality **2.7.8** that it is also positive semi-definite. This last property implies a number of restrictions upon the values of the product moments: in fact, that all the principal minors of U are non-negative. We shall give a direct proof of this result.

Theorem 2.7.1

If U is the product moment matrix of $X_1, X_2, \ldots, X_m$, then $|U| \geqslant 0$, with equality iff there is a relation of the type

$$\sum_1^m c_j X_j \overset{m.s.}{=} 0,$$ **2.7.9**

with not all c_j zero. In particular.

$$|E(X_1 X_2)|^2 \leqslant E(X_1^2)E(X_2^2),$$ **2.7.10**

(Cauchy's inequality) with equality iff there is a relation **2.7.9** *for $m = 2$.*

The second part is just a statement of the first for $m = 2$, but Cauchy's inequality **2.7.10** has a special importance and is much used.

Let us write $|U| = D_m$, to emphasize the dependence upon m. We shall establish the result by an induction upon m. For $m = 1$ the statement is that $E(X_1^2) \geqslant 0$, with equality iff $X_1 \overset{\text{m.s.}}{=} 0$, which is evidently true. Suppose then that $D_r \geqslant 0, (r < m)$. Let

$$\Delta_m = \min_{c_1, c_2, \ldots, c_{m-1}} E\left(\sum_1^{m-1} c_k X_k + X_m\right)^2 \qquad \text{2.7.11}$$

so that $-\sum_1^{m-1} \hat{c}_k X_k$ is the least square approximant of X_m in terms of $X_1, X_2, \ldots, X_{m-1}$ if the $\hat{c}_k$ are the minimizing values in equation **2.7.11**. We have then (cf. equation **2.6.3**)

$$\sum_{k=1}^m u_{jk}\hat{c}_k + u_{jm} = 0 \quad (j = 1, 2, \ldots, m-1) \qquad \text{2.7.12}$$

and, in virtue of equations **2.7.11** and **2.7.12**,

$$\Delta_m = u_{mm} - \sum_{k=1}^m u_{mk}\hat{c}_k. \qquad \text{2.7.13}$$

Eliminating the $\hat{c}_k$ from equations **2.7.12** and **2.7.13** we find that

$$D_m = \Delta_m D_{m-1}, \qquad \text{2.7.14}$$

and, since Δ_m is non-negative, we must have $D_m \geqslant 0$, so the induction is complete. We shall have $D_m = 0$ iff either $\Delta_m = 0$ or $D_{m-1} = 0$. The first case holds iff $\sum \hat{c}_k X_k + X_m \overset{\text{m.s.}}{=} 0$, and so the second holds iff an analogous relation is valid for some lower value of m. The theorem is thus proved.

Note that the relation

$$E(X^2) \geqslant [E(X)]^2, \qquad \text{2.7.15}$$

evident from equation **2.7.5**, is a particular case of Cauchy's inequality.

Exercises

2.7.5   Carry through the proof of the theorem explicitly in the case $m = 2$, to obtain Cauchy's inequality.

2.7.6   Recall the covariance, defined in exercise 2.6.4. Note that

$$\text{cov}(X, X) = \text{var}(X).$$

Show that

$$[\text{cov}(X, Y)]^2 \leqslant \text{var}(X)\text{var}(Y). \qquad \text{2.7.16}$$

Show also that if $\text{var}(X) > 0$, then equality can hold in **2.7.16** iff there is an inhomogeneous linear relationship of the form

$$Y \overset{\text{m.s.}}{=} \alpha + \beta X.$$

2.7.7 Note that

$$\text{cov}(aX, bY) = ab\,\text{cov}(X, Y), \qquad \text{var}(aX) = a^2\text{var}(X),$$

where a, b are constants.

2.7.8 The jth moment of X is defined as

$$\mu_j = E(X^j).$$

Show that

$$\begin{vmatrix} 1 & \mu_1 \\ \mu_1 & \mu_2 \end{vmatrix} \geq 0 \qquad \begin{vmatrix} 1 & \mu_1 & \mu_2 \\ \mu_1 & \mu_2 & \mu_3 \\ \mu_2 & \mu_3 & \mu_4 \end{vmatrix} \geq 0$$

and generalize.

2.7.9 *Jensen's inequality.* A *convex function* $\phi(x)$ is one for which

$$\phi(px + qx') \leq p\phi(x) + q\phi(x'),$$

where p, q are non-negative and add to unity. From this it follows that the derivative $\phi'(x)$ is increasing with x (if it exists) and that for any fixed x_0 there exists a constant α such that

$$\phi(x) \geq \phi(x_0) + \alpha(x - x_0)$$

(identifiable with $\phi'(x_0)$, if this exists). Show that

$$E[\phi(X)] \geq \phi[E(X)].$$

2.7.10 Since $|X|^r$ is a convex function of X for $r > 1$, we have

$$E(|X|^r) \geq [E(|X|)]^r$$

by Jensen's inequality. Infer from this that $[E(|X|^r)]^{1/r}$ is an increasing function of r $(r \geq 0)$.

3 Probability

3.1 Probability measure

In section 2.3 we introduced the idea of an event as a set A of points in Ω, and the probability of this event as $P(A) = E(I_A)$. Since I_A is the r.v. 'proportion of times A has occurred' (which can only be 1 or 0, in a single experiment), $P(A)$ can be characterized as the *expected proportion of times that A occurs*.

A number of properties of $P(A)$ follow immediately from the axioms.

Theorem 3.1.1

The probability $P(A)$ has the properties

(i) $0 \leqslant P(A) \leqslant 1.$ **3.1.1**

(ii) $P(\Omega) = 1.$ **3.1.2**

(iii) *If A and B are mutually exclusive events (disjoint sets) then*

$$P(A + B) = P(A) + P(B).$$ **3.1.3**

(iv) *If $\{A_n\}$ is a sequence increasing to A then*

$$P(A) = \lim_n P(A_n).$$ **3.1.4**

Assertions (i) and (ii) follow from the relations $0 \leqslant I_A \leqslant 1$, and $I_\Omega = 1$. Assertion (iii) follows from the relation

$$I_{A+B} = I_A + I_B,$$ **3.1.5**

where $A + B$ is the set of all points contained in either A or B: this formula is valid only if A and B have no overlap, i.e. are disjoint. By saying that $\{A_n\}$ is increasing we mean that if ω belongs to A_n then it also belongs to A_{n+1}. That is, the set A_n is contained in A_{n+1}, this is written as $A_n \subset A_{n+1}$. Assertion (iv) then follows if we apply Axiom 5 of section 2.2 to the monotone increasing sequence of r.v.s $\{I_{A_n}\}$.

Although formally immediate, these results deserve amplification. For example, we have introduced the important concept of the event, or set, $A + B$ in a rather casual fashion.

Equation **3.1.2** states that the set Ω of all possible outcomes is the *certain event*, and has probability one. Correspondingly, $P(\emptyset) = 0$, where $\emptyset$ is the empty set (the impossible event).

If A and B are two sets (events), then in applications we soon find ourselves considering the derived sets (events) AB and $A\cup B$. The former is the set of all points belonging to both A and B (the event that both A and B occur), and the latter is the set of all points belonging to at least one of A or B (the event that at least one of A or B occurs).

In the particular case $AB = \varnothing$ the sets A and B are said to be *disjoint* (the events are *mutually exclusive*), and in this case it is conventional to write $A\cup B$ as $A+B$. Moreover, if we use the notation $A+B$, then it will be understood that A and B are disjoint.

For such events the vital relation **3.1.3** holds; it is often referred to as the 'additive law of probability'. We see from relations **3.1.5** and **3.1.3** that I_A and $P(A)$ are both *additive set functions*, i.e. functions which attach a number to a set, this number for a union of disjoint sets being just the sum of the numbers for the individual sets.

Let us now go back to the football example of section 1.5, in the case where we took the coordinates ω of Ω as being the scores (s_1, s_2) obtained by the home and visiting teams respectively. If A is the event 'the home team won' and B is the event 'a draw' then these correspond to the sets $\{s_1 > s_2\}$ and $\{s_1 = s_2\}$ Thus $AB = \varnothing$, and relation **3.1.3** holds; that is

$$P(\text{win or draw}) = P(\text{win}) + P(\text{draw}).$$

On the other hand, the events 'the home team scores' and 'the visiting team scores' are not mutually exclusive, and relations **3.1.5** and **3.1.3** would not hold for these events. In fact, seeing that $I_A + I_B$ counts double those cases where both teams score, we have

$$P(\text{score}) \leqslant P(\text{home score}) + P(\text{visiting score}).$$

If $\{A_n\}$ is an increasing sequence, so that $A_n \subset A_{n+1}$, then occurrence of the event A_n implies occurrence of the event A_{n+1}. One has to go to infinite sample spaces to obtain a non-trivial example of such a sequence. For instance, A_n might be the event that a population becomes extinct by the nth generation; obviously A_n then implies A_{n+1}. One would interpret the limit event A as 'ultimate extinction', with probability determined by relation **3.1.4**. The sample space is certainly infinite, since one cannot guarantee that extinction will have occurred within a given time.

The fact that $P(A)$ is a positive additive set function characterizes it as a *measure* on the sets of points in Ω, and it is known as the *probability measure*. The usual approach is to take the assertions of the theorem as the axioms for such a measure (relation **3.1.1** can be weakened to $P(A) \geqslant 0$ for this purpose, since the property $P(A) \leqslant 1$ can then be deduced) and to found a theory of probability on these axioms. For reasons which we shall partly explain in the next section, we have instead axiomatized the concept of expectation, so that relations **3.1.1–4** are deductions rather than postulates. However, the two approaches are of course mutually consistent, and in the end one covers much the same material: the difference at the introductory level is largely one of order and emphasis.

Just as we had to admit the idea that $E(X)$ might be known only for certain Xs, with the possibility of extension to other r.v.s by use of the axioms, so $P(A)$ may be known only for certain As, with the possibility that the axioms again provide a means of extension. Indeed, it is customary to restrict the sets A of Ω to which a probability measure is assigned at all, in order to be sure that it can be assigned consistently with the axioms. These points are discussed further in sections 3.5 and 11.1.

Exercises

3.1.1 Note that, if some experimental outcomes have zero probability, then we will have $P(A) = 0$ or $P(A) = 1$ for sets A which are respectively larger than $\varnothing$ or smaller than Ω.

3.1.2 Let $N(A)$ be the number of molecules in a region A of a gas contained in a total region Ω. Then $N(A)$ is an example of an additive set function. Find others. If $\mathscr{E}(A)$ is the total energy of the molecules in A, then $\mathscr{E}(A)$ is additive if and only if there is no interaction between molecules.

3.2 Probability and expectation

The additive law **3.1.3** extends immediately to give

$$P\left(\sum_{1}^{n} A_j\right) = \sum_{1}^{n} P(A_j),\qquad\qquad\qquad \textbf{3.2.1}$$

where by use of the notation $\sum_{1}^{n} A_j$ we imply that the A_j are disjoint, and $\sum_{1}^{n} A_j$ is then the set of all points contained in $A_1, A_2, \ldots, A_n$ jointly.

Suppose, for example, we were observing wind-speed v in miles per hour, and A_j were the event $j-1 \leqslant v < j$. These events are certainly mutually exclusive, and $\sum_{1}^{n} A_j$ would be the event 'wind-speed is less than n m.p.h.', with probability $\sum_{1}^{n} P(A_j)$.

In fact, by relation **3.1.4**, the additivity relation **3.2.1** holds even if n is infinite: this is known as the property of *countable additivity*, or *σ-additivity*. So, for the wind-speed example, $\sum_{1}^{\infty} A_j$ would be the event $0 \leqslant v < \infty$, that 'wind-speed is finite,' and this would have probability $\sum_{1}^{\infty} P(A_j)$.

Suppose that

$$\sum_{1}^{n} A_j = \Omega.\qquad\qquad\qquad \textbf{3.2.2}$$

This implies that the events $\{A_j\}$ are not merely mutually exclusive, but also *exhaustive*; between them they include all possibilities, and one of them must happen. (From the set theory point of view, the collection of sets $\{A_j\}$ is said to constitute a *decomposition* of Ω.)

For instance, in the football example we could take A_1, A_2, A_3 as 'win', 'lose' and 'draw' respectively: these are certainly exclusive and exhaustive, and constitute a decomposition. For the wind-speed example, if infinite wind-speed is known to be physically impossible, so that it can be excluded as a conceivable experimental outcome, then the finite half-axis $0 \leqslant v < \infty$ can be taken as the sample space Ω, and A_1, A_2, ... constitute a decomposition of it.

Such a decomposition in fact corresponds to a coarser reporting of the experiment, in which A_1, A_2, ... can be taken as the points of a new and simpler sample space Ω_A. For the football example, to report one of the outcomes A_1, A_2 or A_3 means reporting 'win, lose or draw' instead of the actual score. For the wind example, reporting of one of A_1 A_2, ... means reading wind-speed to the nearest integral number of m.p.h. below, instead of reporting the exact speed.

Suppose we consider a r.v. $X(\omega)$ observable on Ω_A; that is, $X(\omega)$ takes the same value (say x_j) for any ω in A_j. For example, one might have had a bet of £1 on the result of the football match, so that if $X(\omega)$ is the amount won on the bet, then

$$X(\omega) = \begin{cases} 1 & (\omega \in A_1; \text{'win'}), \\ 0 & (\omega \in A_2; \text{'draw'}), \\ -1 & (\omega \in A_3; \text{'lose'}). \end{cases}$$

So, looking on $X(\omega)$ as a function of the scores s_1, s_2, it takes the value $+1$ for all $\omega \in A_1$, i.e. for all s_1, s_2 such that $s_1 > s_2$.

Note then that we can write

$$X(\omega) = \sum_j x_j I_{A_j}(\omega). \qquad \textbf{3.2.3}$$

Taking expectations in this equation, we obtain

$$E(X) = \sum x_j P(A_j). \qquad \textbf{3.2.4}$$

This is certainly a legitimate procedure if the sum **3.2.4** is absolutely convergent.

Thus for the particular r.v.s which are observable on Ω_A, the expectation is given in terms of the probability measure by the sum **3.2.4**: this is a partial converse of the relation $P(A) = E(I_A)$, defining P in terms of E.

Consider now decompositions of Ω which are finer and finer, until at last we reach the finest permitted, in which the sets A_j correspond to individual ω-values, i.e. to fully-specified experimental outcomes. Then the sum **3.2.4** will have a limit form which we shall write

$$E(X) = \int X(\omega)P(d\omega) \qquad \qquad \textbf{3.2.5}$$

for an X observable on Ω. This is the general representation of $E(X)$ as an integral with respect to P-measure.

However, there are many technicalities and pitfalls on the path from **3.2.4** to **3.2.5**. Essentially, one is trying to form a very general type of integral over a very general space Ω. One must demonstrate that the integral thus obtained is valid for all Xs of interest, that its value is independent of the sequence of decompositions used etc.

It is partly for this reason that we took the concept of expectation as primitive, rather than that of probability. In this way one avoids the difficulty that a representation of E in terms of P may need a very general concept of integral. Even at a more elementary level there are advantages; for example, discrete and continuous distributions need not be introduced separately, with all concepts for the two cases developed in semi-parallel. By working from the expectation one can quickly see both as special cases of a probability process, necessarily sharing many properties. And finally, there is the operational justification of reality. It is often the expectations of particular r.v.s that are given, rather than a probability distribution (as in the least square situation of section 2.6); it is often the expectation of a particular r.v. that is of interest (as in the optimization problem of section 2.5); and it is an expectation rather than a distribution which is 'estimable', by the formation of long-term averages over repeated experiments.

Nevertheless, the representation **3.2.5** is, of course, basic, and has very great interest. Its proof in a general case (the *Riesz representation*) is a matter of considerable sophistication. In the next section (which can be omitted in a first reading) we present a more limited result in this direction, which is nevertheless interesting in that it does not appeal to the idea of increasingly fine decompositions of Ω. Such limit arguments, although useful and graphic, are often difficult to justify.

Exercises

3.2.1 Show that the condition $\sum_{1}^{n} I_{A_j}(\omega) \equiv 1$ $(\omega \in \Omega)$ is equivalent to **3.2.2**, that is, it implies that events A_j are both mutually exclusive and exhaustive.

3.2.2 Suppose that X can only assume the values $0, 1, 2, \dots$. By using indicator functions, or otherwise, show that

$$\sum_{n=0}^{\infty} P(X > n) = E(X).$$

3.3 Expectation as an integral

[Optional reading]

Suppose that the values $E(X_j)$ are given for a number of random variables $X_j (j = 1, 2, \ldots, m)$ consistently with the axioms. We wish then to show that there exists a measure P such that the m expectations are *simultaneously* representable in the integral form

$$E(X_j) = \int X_j(\omega)P(d\omega) \quad (j = 1, 2, \ldots, m). \qquad \textbf{3.3.1}$$

Nevertheless, we wish to avoid the construction of these integrals explicitly from limits of sums such as **3.2.4**, and so must find an alternative characterization of such an integral.

We regard the integral $\int X(\omega)P(d\omega)$ as a functional of the r.v. $X(\omega)$, which depends upon the individual measure P; and denote it by $I(X, P)$. We shall accept $X(\omega_0)$ as a particular case of such an integral, and say that it corresponds to the case of a measure (or distribution) completely concentrated on the value ω_0 $(\omega_0 \in \Omega)$. We can generate further measures from these by averaging: we shall say that if P and P' are measures, then so is $P'' = pP + qP'$, with $I(X, P'')$ defined as being equal to $p I(X, P) + q I(X, P')$. Here p and q are non-negative values adding to unity. Let $\mathscr{P}(\Omega)$ denote the smallest class of measures obtainable in this way: i.e. the class is closed under averaging, and contains the one-point distributions on Ω. (This is in fact the class of distributions of finite support on Ω.)

Theorem 3.3.1

Suppose the values $E(X_j) (j = 1, 2, \ldots, m)$ have been prescribed consistently with the axioms of section 2.2 for r.v.s defined on Ω. Then there exists at least one probability measure P in $\mathscr{P}(\Omega)$ such that

$$E(X_j) = I(X_j, P),$$

for $j = 1, 2, \ldots, m$ simultaneously.

For, consider an m-dimensional space S with vector coordinate ξ, where

$$\xi_j = E(X_j) \quad (j = 1, 2, \ldots, m). \qquad \textbf{3.3.2}$$

Consider the ξ-region in S with coordinates determined by equation **3.3.2**, as P varies over all distributions of $\mathscr{P}(\Omega)$; denote this by D, as in Figure 2. If ξ and ξ' lie in D, then so does $\xi'' = p\xi + q\xi'$. That is, D is convex. (See the note at the end of the section.) Furthermore it contains the points with coordinate $\xi_j = X_j(\omega) (j = 1, 2, \ldots, m)$, for all $\omega \in \Omega$.

The assertion of the theorem is that, if α_j is the prescribed value of $E(X_j)$, then the point $\xi = \alpha$ lies in D. Suppose it does not. Then there exists a hyperplane (see note)

$$\sum \lambda_j \xi_j + \mu = 0 \qquad \textbf{3.3.3}$$

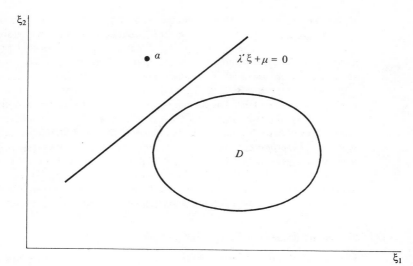

Figure 2

strictly separating α and D, i.e. such that $\sum \lambda_j \xi_j + \mu$ has strictly opposite signs for $\xi = \alpha$ and $\xi \in D$ (say, negative and positive respectively). Consider the r.v. $Y = \sum \lambda_j X_j + \mu$. Since the point with coordinates $X_j(\omega)$ belongs to D we have

$$Y(\omega) > 0 \quad (\omega \in \Omega), \qquad\qquad 3.3.4$$

while $\quad E(Y) = \sum \lambda_j \alpha_j + \mu < 0.$ $\qquad\qquad 3.3.5$

But relations **3.3.4** and **3.3.5** are contradictory, so our supposition, that α does not lie in D, is incorrect, and the theorem is established.

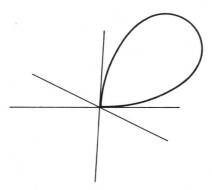

Figure 3 A convex set with more than one supporting hyperplane at a boundary point

A note on convexity. Although simple, this is one of the most important concepts in mathematics, and permeates probability theory. A region D is said to be *convex* if the fact that ξ and ξ' belong to D implies that $p\xi + q\xi'$ also belongs to D, for p and q non-negative and adding to unity. We take it as geometrically obvious that, if α is a point outside D, then there is a *separating hyperplane*, i.e. a plane **3.3.3** such that the point α lies to one side of it, and all points of D to the other. The formal proof is quite short (see Appendix B of Karlin, 1959, for a concise treatment). A special case of the result, obtained as α converges towards a point on the boundary of D, is that there is at least one *supporting hyperplane* at every point ξ of the boundary of D; that is, a hyperplane which meets D at ξ, but such that all points of D lie either on the hyperplane or to one side of it. This might be called a 'tangent hyperplane', but the case illustrated in Figure 3 indicates that there may be more than one supporting hyperplane at a point, and the word 'tangent' is then less appropriate.

3.4   Elementary properties of events and probabilities

The characterization of events as point sets A in Ω, and of probabilities as a measure $P(A)$ on these sets, implies a large number of simple formal properties, which we shall treat rather rapidly.

Let us first make a tabular summary of a number of equivalent concepts for events and sets (p. 51).

The operations of complementation (**3.4.4**), union (**3.4.5** and **3.4.11**) and intersection (**3.4.6** and **3.4.12**) are important. As in previous sections, we shall use the notation $\sum_1^n A_j$ instead of $\bigcup_1^n A_j$ if the A_j are disjoint, and only in this case.

These derived events are certainly important in practice. For example, suppose all n units of a machine must function if the machine is to function. If A_j is the event 'jth unit fails', then the event 'machine fails' is $\bigcup_1^n A_j$. The event 'jth unit functions' is $\bar{A}_j$, and the event 'machine functions' is $\bigcap_1^n \bar{A}_j$. From this example we perceive the important general identity:

$$\overline{\bigcup_1^n A_j} = \bigcap_1^n \bar{A}_j. \qquad\qquad \textbf{3.4.1}$$

Exercises

3.4.1   The symmetric difference $A \triangle B$ is the set of all ω belonging to one of A or B, but not both. Interpret the event.

3.4.2   Note that $\bar{A}$ is characterized by $A \cup \bar{A} = \Omega$, $A\bar{A} = \varnothing$.

3.4.3   Note that $AB = A$ if $A \subset B$. In particular, $A^2 = A$.

Notation	Set interpretation	Event interpretation
3.4.2 ω	A basic element (or *atom*) of Ω	The *elementary event*: a fully described experimental outcome; the maximal observable
3.4.3 A	The set of elements ω constituting A	The *event A*, made up of a set of experimental outcomes
3.4.4 $\bar{A}$	The *complement* of A; all ω not belonging to A	The event that A does *not* happen
3.4.5 $A \cup B$	The *union* of A and B: all ω belonging to at least one of A or B	The event that *at least one* of A or B happens
3.4.6 $A \cap B$ or AB	The *intersection* of A and B: all ω belonging to both A and B	The event that A and B *both* happen
3.4.7 $\varnothing$	The *empty set*, containing no elements	The *impossible* event
3.4.8 Ω	The whole set, containing all elements	The whole sample space: the *certain* event
3.4.9 $A \subset B$	*Set inclusion*: all ω belonging to A also belong to B	Implication of events: if A occurs, so must B
3.4.10 $AB = \varnothing$	The sets A and B are *disjoint*: they have no common element	The events A and B are *mutually exclusive*: they cannot happen simultaneously
3.4.11 $\bigcup_{j=1}^{n} A_j$	The *union* of sets $A_1, A_2, \ldots, A_n$: the set of ω belonging to at least one of these	The event that at least one of $A_1, A_2, \ldots, A_n$ happens
3.4.12 $\bigcap_{j=1}^{n} A_j$	The *intersection* of sets $A_1, A_2, \ldots, A_n$: the set of ω belonging to all of these.	The event that $A_1, A_2, \ldots, A_n$ all happen.

3.4.4   Show that $A \cup B = A + \bar{A}B$. Generalize.

3.4.5 A_{jk} is the event 'kth unit of jth machine functions' ($j = 1, 2, \ldots, m$; $k = 1, 2, \ldots, m$). Write symbolically the event 'at least one machine functions'.

3.4.6   Write symbolically the event 'at least two of $A_1, A_2, \ldots, A_n$ occur'.

The rules relating sets manifest themselves in the rules relating their indicator functions. We have, of course,

$$I_\varnothing = 0 \qquad I_\Omega = 1 \qquad I_A + I_{\bar{A}} = 1. \tag{3.4.13}$$

The basic rule is

$$I_{AB} = I_A I_B \tag{3.4.14}$$

with an immediate generalization to the case of $\bigcap_1^n A_j$. From relations **3.4.1** and **3.4.14** we see that

$$I_{\cup A_j} = 1 - \prod (1 - I_{A_j}). \tag{3.4.15}$$

In particular

$$I_{A \cup B} = I_A + I_B - I_{AB}. \tag{3.4.16}$$

From relation **3.4.15** we deduce that, for the case of disjoint sets,

$$I_{\Sigma A_j} = \sum I_{A_j}. \tag{3.4.17}$$

Formulae **3.4.16** and **3.4.17** generalize the relation **3.1.6**, which we have already seen to be basic.

The inequalities

$$I_{\cup A_j} \leqslant \sum I_{A_j}, \tag{3.4.18}$$

$$I_{\cap A_j} \geqslant 1 - \sum I_{\bar{A}_j} \tag{3.4.19}$$

are often useful.

Exercises

3.4.7   Prove the inequalities **3.4.18** and **3.4.19** formally.

3.4.8   Show that $I_{A \triangle B} = (I_A - I_B)^2$.

3.4.9   Under what circumstances is it true that

$$I_{\cup A_j} = \sum I_{A_j} - \sum \sum_{j < k} I_{A_j A_k}?$$

By taking expectations of these relations we obtain a number of probability identities immediately. From relation **3.4.13** we have

$$P(\bar{A}) = 1 - P(A). \tag{3.4.20}$$

From **3.4.17** we obtain **3.2.1** and from **3.4.16**,

$$P(A \cup B) = P(A) + P(B) - P(AB).$$ **3.4.21**

From **3.4.1** and **3.4.18** we see that

$$P\left(\bigcup_1^n A_j\right) = 1 - P\left(\bigcap_1^n \bar{A}_j\right) \leqslant \sum_1^n P(A_j)$$ **3.4.22**

$$P\left(\bigcap_1^n A_j\right) \geqslant 1 - \sum_1^n P(\bar{A}_j).$$ **3.4.23**

The inequalities, which are very useful, are known as *Boole's inequalities*. If we expand the product in relation **3.4.15** we obtain

$$P\left(\bigcup_1^n A_j\right) = \sum P(A_j) - \sum_{j<k}\sum P(A_j A_k) + \sum_{j<k<l}\sum\sum P(A_j A_k A_l) - \ldots,$$ **3.4.24**

and the expressions obtained by breaking off this expansion from a certain point onwards (e.g. after the simple sum, or after the double sum) actually provide alternating upper and lower bounds for $P\left(\bigcup_1^n A_j\right)$.

Exercises

3.4.10 Prove the assertion just made.

3.4.11 Show that

$$P\left(\bigcup_1^n A_j\right) = P(A_1) + P(\bar{A}_1 A_2) + P(\bar{A}_1 \bar{A}_2 A_3) + \ldots.$$

What would be the corresponding identity in indicator functions?

3.4.12 By minimizing the left-hand member of the inequality

$$E(I_{\cup A_j} - \sum c_j I_{A_j})^2 \geqslant 0$$

with respect to the coefficients c_j, show that

$$P\left(\bigcup_1^n A_j\right) \geqslant \Pi_1' \Pi_2^{-1} \Pi_1,$$

where Π_1 is the vector with elements $P(A_j)$, and Π_2 the matrix with elements $P(A_j A_k)$. This is useful in the study of maxima of r.v.s: if A_j is the event $X_j \geqslant x$, then $\bigcup_1^n A_j$ is the event $\max_j X_j \geqslant x$.

3.5 **Fields of events**

Given expectations of r.v.s X_j, we can evaluate those of r.v.s $\sum c_j X_j$. However, given probabilities for events A_j, we can really only calculate those of complementary events $\bar{A}_j$, or of unions $\sum_j A_{n_j}$ of such groups of disjoint sets as happen to exist among the A_j. If it happens that $A_1 \subset A_2$ then we can also calculate P for $A_2 - A_1 = A_2 \bar{A}_1$, for, by the additive law,

$$P(A_1) + P(A_2 - A_1) = P(A_2). \qquad \text{3.5.1}$$

One useful situation is that in which the family $\{A_j\}$ constitutes a decomposition of Ω, because then distributions and expectations are determinable for r.v.s observable on Ω_A.

A situation reducible to this one is that for which

$$A_1 \subset A_2 \subset A_3 \subset \ldots \subset A_n = \Omega,$$

for we can then calculate probabilities for

$$B_1 = A_1, \qquad B_j = A_j - A_{j-1} \quad (j = 2, 3, \ldots, n)$$

and $\{B_1, B_2, \ldots, B_n\}$ constitutes a decomposition of Ω.

More generally, if $P(A_1), P(A_2), \ldots, P(A_n)$ are known, and if it is true that $P(B), P(C)$ known implies $P(BC)$ known (the intersection assumption) then we can determine probabilities for $A_{1j_1} A_{2j_2} \ldots A_{nj_n}$ $(j_1, j_2, \ldots, j_n = 0, 1)$, where $A_{j_0} = A_j, A_{j_1} = A_j$. These 2^n sets constitute a decomposition: the *decomposition generated by* $\{\bar{A}_j\}$.

The intersection assumption is probably not a very realistic one, save in the simple cases quoted above, but it does provide a convenient approach in cases where one has to deal with a continuous infinity of sets.

For example, suppose that the sets of known probability are

$$A(x) = \{\omega : X(\omega) \leqslant x\},$$

for all real x. That is, for a r.v. X one knows the *distribution function*

$$F(x) = P(X \leqslant x) \quad (-\infty < x < +\infty). \qquad \text{3.5.2}$$

Then $A(x)$ is monotone increasing in x, and this is the continuous analogue of the case $A_1 \subset A_2 \subset \ldots$ considered above. We have

$$P(x - h < X \leqslant x) = F(x) - F(x - h), \qquad \text{3.5.3}$$

for arbitrarily small positive h, and, in a certain sense, one can attain a decomposition of Ω in which the sets are $X(\omega) = x$ $(-\infty < x < +\infty)$. However, in most cases these sets will individually have zero probability, and to build up other sets from them will require a non-denumerable infinity of operations. In general, to be sure of what we are doing, we must confine our attention to sets which can be reached from the original ones $A(x)$ in at most countably many operations.

Consider then, the family of sets $\mathscr{X}$ obtained by applying countably many intersection and complementation (and hence union) operations to the sets $A(x)$. A family of sets which contains Ω and is closed under the operations of complementation and intersection is called a *field*; a family closed under countably many such operations is called a *σ-field*. The family $\mathscr{X}$ is both of these: it is termed the *Borel field of sets generated from* $\{A(x)\}$.

It will be shown in section 11.1 that, if probabilities are assigned consistently on a given field of sets $\mathscr{B}$ and one assumes the intersection property, then probabilities are also determined, and consistently, for sets in the Borel field generated from $\mathscr{B}$. But this result is a very special case of a more general one (Theorem 10.3.1) concerning families of expectations, which is no more difficult to enunciate and prove, and is probably more natural in its setting.

Some authors (e.g. Feller, 1966) reserve the term 'random variable' for observables $X(\omega)$ whose expectation is determinable from the probability measure on some given field $\mathscr{B}$ of subsets of Ω. We shall adhere to the wider definition of section 2.1, referring to this second concept as a 'r.v. measurable on $\mathscr{B}$'.

Exercise

3.5.1 Note that, if $\{A_j\}$ constitutes a countable decomposition of Ω, then the Borel field generated from it is just $\{A_j\}$ itself.

4 Some Simple Processes

4.1 Equiprobable sample spaces

In order to start upon a problem one must be able to specify, on physical grounds, the values of a number of expectations – perhaps these may be probabilities. In this chapter we shall consider the simplest case of all: that in which the sample space contains a finite number K of points (that is, where there are just K possible outcomes) and where symmetry arguments make it plausible to assume that these K elementary events are all equally probable.

Quite a number of interesting problems can be treated in this way. Some nineteenth-century approaches to probability postulated that such an underlying set of equiprobable elementary events could always be found; but this basis was found too limited. It is of interest, nevertheless, to note that modern physics concerns itself more and more with quantized variables (i.e. discrete systems) and that the idea of equiprobable states is fundamental in the statistical mechanics of equilibrium processes (see section 6.5 and exercise 12.1.7).

The probability of an elementary event ω_k $(k = 1, 2, \ldots, K)$ must be $1/K$ One way to specify the process is to say that the expectation functional is

$$E(X) = \frac{1}{K} \sum_{k=1}^{K} X(\omega_k). \qquad \textbf{4.1.1}$$

In particular, the probability of an event A is

$$P(A) = \frac{\text{number of elementary events in } A}{K}. \qquad \textbf{4.1.2}$$

The calculation of probabilities is thus just a matter of counting the number of elementary events whose occurrence implies the occurrence of A. Sometimes this counting task is trivial, but sometimes it is a formidable combinatorial problem, requiring advanced methods for its solution. In a sense these purely technical challenges have nothing to do with the theory of probability as such. However, one cannot form an idea of either the spirit of the subject or of the range of its application unless one works through some of these cases.

The coin-tossing and card-drawing examples of this chapter have a regrettably artificial character. They do constitute, however, simple and convincing examples of a process with an equiprobable sample space. Genuinely physical examples usually carry complications which, for present purposes, are irrelevant.

4.2 A coin-tossing example; stochastic convergence

Suppose that one tosses a coin n times, observing the complete sequence of heads and tails thus obtained. There are 2^n such sequences, and consequently $K = 2^n$ elementary events. We shall assume that all these sequences are equally likely; that is, a head is as likely to appear as a tail at any given toss, and the order of a sequence does not affect its probability. This assumption seems reasonable; it amounts in fact to two assumptions: that the coin is fair, and that the results of different tosses are statistically independent. We have not yet discussed statistical independence (see section 5.5) but the reader may have an intuitive feeling for the idea.

Let ξ_j be a variable indexing the results of the jth toss, and so taking the values 'head' and 'tail'. Then $\omega = (\xi_1, \xi_2, \ldots, \xi_n)$ is the coordinate of the sample space, and formula **4.1.1** becomes

$$E(X) = 2^{-n} \sum_{\xi_1} \sum_{\xi_2} \cdots \sum_{\xi_n} X(\xi_1, \xi_2, \ldots, \xi_n). \tag{4.2.1}$$

One r.v. defined on this experiment is of interest: the number of heads in the sequence, which we shall denote by R. Consider the event $R = r$. Of the 2^n sequences, just

$$\binom{n}{r} = \frac{n!}{r!(n-r)!}$$

(the number of ways in which r indistinguishable objects and another $n-r$ indistinguishable objects can be ordered) will contain r heads. Thus, by formula **4.1.2**, we have

$$P_r = P(R = r) = \binom{n}{r} 2^{-n} \quad (r = 0, 1, 2, \ldots, n). \tag{4.2.2}$$

The probabilities P_r describe the probability distribution of the r.v. R. The distribution is a symmetric one about the value $r = \frac{1}{2}n$, with a strong peak around this value: in Figure 4 we graph it for the case $n = 15$.

We find from formula **4.2.1** that

$$E(R) = \sum_r r P_r = \frac{n}{2}, \tag{4.2.3}$$

$$E(R^2) = \sum_r r^2 P_r = \frac{n^2 + n}{4}, \tag{4.2.4}$$

so that

$$\mathrm{var}(R) = \frac{n}{4} \tag{4.2.5}$$

(see equation **2.7.5**). The reader may not see immediately how the sums **4.2.3** and **4.2.4** have been evaluated. We shall defer this point for the moment,

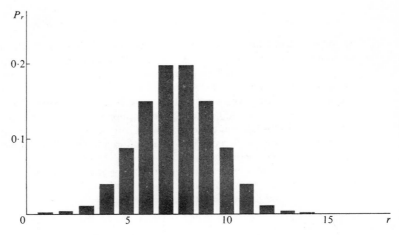

Figure 4 The binomial distribution for $n = 15$, $p = \frac{1}{2}$

because it will be covered easily in an alternative approach to the problem given later in the section.

If we consider the *proportion* of heads observed, then we see from relations **4.2.3** and **4.2.5** that

$$E\left(\frac{R}{n}\right) = \frac{1}{2}, \qquad \textbf{4.2.6}$$

$$E\left(\frac{R}{n} - \frac{1}{2}\right)^2 = \operatorname{var}\left(\frac{R}{n}\right) = \frac{1}{n^2}\operatorname{var}(R) = \frac{1}{4n}. \qquad \textbf{4.2.7}$$

Both of these results are gratifying. The expected proportion of heads is $\frac{1}{2}$; just what we should hope for if the coin was a fair one. Furthermore, the mean squared deviation **4.2.7** tends to zero as n becomes large. Extending the notation of equation **2.2.7**, we could say that $R/n \xrightarrow{\text{m.s.}} \frac{1}{2}$. Applying Chebichev's inequality **2.7.6**, we find that this has the implication

$$P\left(\left|\frac{R}{n} - \frac{1}{2}\right| > \varepsilon\right) \leqslant \frac{1}{4n\varepsilon^2} \to 0, \qquad \textbf{4.2.8}$$

so that the probability that R/n deviates from $\frac{1}{2}$ by more than an arbitrarily small amount tends to zero with increasing n.

But this is just the kind of convergence remarked upon in section 1.2; that a sample average should tend in some sense to a limit value (the 'expected value') with increasing sample size. The proportion R/n is indeed a sample average, for

$$R = \sum_{j=1}^{n} I(\xi_j), \qquad \textbf{4.2.9}$$

where $I(\xi)$ is 1 or 0 according as the toss results in a head or a tail. The mode of convergence is now made quite explicit by **4.2.7** and its implication **4.2.8**. Furthermore, the convergence has been *deduced* from our axioms, rather than observed. This is encouraging: that our theory should reproduce just the kind of limiting behaviour observed in practice, which constituted the empirical basis for the whole study.

The types of convergence exemplified by **4.2.7** and **4.2.8** are weaker than ordinary convergence, but are obviously useful adaptations of the idea to the case of a random sequence. They are types of *stochastic convergence*, to which we shall return in sections 6.1, 6.2 and 9.2. The word 'stochastic' (from the Greek *stochos*, a knucklebone) is a synonym for 'random'.

Let us consider another treatment of the problem, which avoids combinatorial or distributional ideas altogether. Consider the r.v. z^R, where z is a constant. We see from relations **4.2.1** and **4.2.9** that

$$
\begin{aligned}
E(z^R) &= 2^{-n} \sum \sum \ldots \sum z^{\sum_j I(\xi_j)} \\
&= \left(\tfrac{1}{2} \sum_\xi z^{I(\xi)} \right)^n \\
&= \left(\frac{1+z}{2} \right)^n.
\end{aligned}
\qquad \textbf{4.2.10}
$$

By the definition of P_r, expectation **4.2.10** must also equal $\sum z^r P_r$, so we should be able to identify P_r with the coefficient of z^r in the expansion of expression **4.2.10** in powers of z. This identification yields formula **4.2.2** immediately, without appeal to combinatorial arguments.

Although z is a constant in that it is not a r.v., we shall regard it as a mathematical variable, and the function

$$
\Pi(z) = E(z^R) \qquad \textbf{4.2.11}
$$

will be termed the *probability generating function* (abbreviated to p.g.f.) of R. Use of the p.g.f., or some analogue of it, constitutes one of the most important techniques of probability theory.

If we differentiate relation **4.2.11** under the expectation sign and set $z = 1$ we obtain

$$
E(R) = \Pi'(1), \qquad \textbf{4.2.12}
$$

so that $E(R)$ can readily be determined if the p.g.f. is known The same operation with a v-fold differentiation yields the relation

$$
E(R^{(v)}) = \Pi^{(v)}(1) \quad (v = 0, 1, 2, \ldots), \qquad \textbf{4.2.13}
$$

where we have used the factorial power

$$
R^{(v)} = R(R-1)(R-2)\ldots(R-v+1)
$$

59   A coin-tossing example; stochastic convergence

and $\Pi^{(v)}$ is the vth differential of Π. So, not only can one determine the mean, but all the expectations **4.2.13**, from which the moments $E(R^v)$ can readily be determined.

Thus, in case **4.2.10** one finds

$$E(R^{(v)}) = 2^{-v}n^{(v)}. \qquad \qquad \textbf{4.2.14}$$

Relation **4.2.14** for $v = 1, 2$ yields the two expectations **4.2.3** and **4.2.4**.

Exercises

4.2.1   A player tosses a coin until a head turns up, but stops in any case after a maximum of n tosses. Let N_1 denote the number of tosses he makes. Determine the distribution of N_1, $E(z^{N_1})$ and $E(N_1)$, under the assumptions of this section. Consider these results in the limit case $n \to \infty$.

4.2.2   Note that relation **4.2.13** is equivalent to the formal expansion

$$\Pi(1+\alpha) = \sum_{v=0}^{\infty} \frac{\alpha^v}{v!} E(R^{(v)}).$$

4.3   A more general example; the binomial distribution

A coin can show only two faces: let us generalize the example of the last section by supposing that a single experiment has m possible outcomes, all equally likely. Suppose, for example, that one draws a card from a pack of m different cards, in such a way that any card is equally likely to be chosen. We could also imagine that one repeats this n times, replacing the card and reshuffling each time. It is then plausible that each of the m^n possible sequences obtainable in this way has the same probability, m^{-n}.

This example is not entirely artificial – one might generate such a sequence in tests of telepathy, for instance. A variant of it would be to suppose that each of n molecules could take any one of m possible positions, and that all m^n possible allocations (the molecules being supposed distinguishable) are equally likely.

As another example (cf. section 1.3) consider a census in which a sample of n is chosen from a population of m people. If this sample is chosen by n repeated 'draws' in which all m possibilities have equal chance, then we have the same situation as above. Of course, since this means that a given person can appear in the sample more than once, a more rational method would be that of sampling 'without replacement' (see section 4.5).

In any case, if ξ_j represents the outcome of the jth draw, then, corresponding to equation **4.2.1**, we have

$$E(X) = m^{-n} \sum_{\xi_1} \sum_{\xi_2} \cdots \sum_{\xi_n} X(\xi_1, \xi_2, \ldots, \xi_n), \qquad \qquad \textbf{4.3.1}$$

where each summation runs over the m possible outcomes of a single draw.

Suppose now that, of the m cards, m_1 are red and m_2 are black $(m_1 + m_2 = m)$, and that we are interested in the number of red cards drawn in the sequence. We shall denote this r.v. by R as in the previous section.

The probability of a fully specified sequence is m^{-n}. If we specify only the colours, and their order, then the probability of a sequence containing r red and $n-r$ black cards will be $m_1^r m_2^{n-r}/m^n$. This follows from equation **4.1.2**, since for each red card in the sequence there is a choice of m_1 possibilities, and for each black card a choice of m_2.

If now we do not even specify the order in which the colours occur, but only the total counts r and $n-r$, then by the same argument as in the previous section we have

$$P_r = P(R = r) = \binom{n}{r} \frac{m_1^r m_2^{n-r}}{m^n}$$

$$= \binom{n}{r} p^r (1-p)^{n-r}, \qquad \textbf{4.3.2}$$

where $\qquad p = \dfrac{m_1}{m}.$

Relation **4.3.2** expresses the probability that the sequence will contain exactly r red cards, in terms of p, the proportion of red cards in the pack.

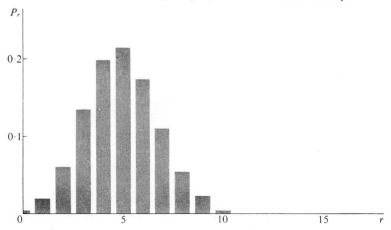

Figure 5 The binomial distribution for $n = 15$, $p = \frac{1}{3}$

Distribution **4.3.2** is known as the *binomial distribution*; we have already encountered it for the special case $p = \frac{1}{2}$ in equation **4.2.2**. It has a maximum near $r = np$, and falls away as r departs from this value in either direction.

The distribution is asymmetric about the value np if $p \neq \frac{1}{2}$, although this asymmetry decreases as n increases. In Figure 5 we graph the distribution for $n = 15$, $p = \frac{1}{3}$.

The binomial distribution is one that occurs repeatedly in practice, for the distribution of quantities such as the number of defective items in a sample of manufactured goods under inspection, the number of people in a sample who fall into a given blood group, etc. We shall give the distribution a rather broader theoretical foundation in section 5.5.

Formula **4.3.2** could also have been derived by the p.g.f. technique. Let $I(\xi)$ denote the r.v. which takes the values 1 or 0 according as the draw is a red card or a black one. Then equation **4.2.9** still holds, and, as in equation **4.2.10**,

$$\Pi(z) = E(z^R) = \left(m^{-1} \sum_\xi z^{I(\xi)} \right)^n$$

$$= \left(\frac{m_1 z + m_2}{m} \right)^n$$

$$= (pz + 1 - p)^n. \tag{4.3.3}$$

Identifying the coefficient of z^r in $\Pi(z)$ with P_r, we obtain expression **4.3.2**. Formula **4.3.3** makes it clear why the distribution is described as 'binomial'.

Applying relation **4.2.13** we find

$$E(R^{(v)}) = n^{(v)} p^v \quad (v = 0, 1, 2, \ldots), \tag{4.3.4}$$

whence

$$E(R) = np, \tag{4.3.5}$$

$$\mathrm{var}(R) = npq. \tag{4.3.6}$$

We have set $1 - p$ equal to q, a standard notation. The mean and variance of the proportion of red cards in the sample are thus

$$E\left(\frac{R}{n} \right) = p, \tag{4.3.7}$$

$$\mathrm{var}\left(\frac{R}{n} \right) = \frac{pq}{n}. \tag{4.3.8}$$

The conclusions of the previous section have a clear generalization: we see from equations **4.3.7** and **4.3.8** that $R/n \xrightarrow{\text{m.s.}} p$, with the implication that

$$P\left[\left| \frac{R}{n} - p \right| > \varepsilon \right] \leqslant \frac{pq}{n\varepsilon^2} \to 0. \tag{4.3.9}$$

Again we observe that the sample proportion converges stochastically with increasing sample size, and to the expectation value p. The result is generalized in exercise 4.3.3.

Exercises

4.3.1   Note from the equation

$$\frac{P_{r+1}}{P_r} = 1 + \frac{(n+1)p - (r+1)}{(r+1)q}$$

that P_r has a single maximum, and that this lies in the range $(np - q, np + p)$.

4.3.2   Suppose that the m cards in the pack are ordered and numbered 1 to m. Show that the probability that all cards drawn fall in the range $1, 2, \ldots, u$ is $(u/m)^n$. Hence show that if U is the r.v. 'largest serial number observed' then

$$P(U = u) = \left(\frac{u}{m}\right)^n - \left(\frac{u-1}{m}\right)^n$$

and that, if m is large, then

$$E(U) \simeq \frac{nm}{n+1} \qquad \text{var}(U) \simeq \frac{nm^2}{(n+1)^2(n+2)}.$$

Hence, if the number of cards in the pack is unknown, one might estimate it from $(n+1)U/n$. This has applications to the observation of registration numbers etc.

4.3.3   *Sampling from a finite population.* Suppose we are sampling a population of m in order to measure a quantitative characteristic, say income. Let the income of the jth member of the population be y_j, and let $\bar{x}$ be the average income in the sample of n taken. Show that for any θ

$$E(e^{\theta\bar{x}}) = \left(m^{-1} \sum_{k=1}^{m} e^{\theta y_k/n}\right)^n,$$

and, hence or otherwise, that

$$E(\bar{x}) = \frac{1}{m} \sum_{1}^{m} y_k = \bar{y}, \text{ say,}$$

and that

$$\text{var}(\bar{x}) = \frac{\sum_{1}^{m} (y_k - \bar{y})^2}{mn}.$$

Thus, $\bar{x} \xrightarrow{\text{m.s.}} \bar{y}$. In words, the sample average converges in mean square to the population average with increasing sample size.

4.4   Multiple classification; the multinomial distribution

Suppose that in the previous example there had been several colours or suits of cards, instead of just two. We shall suppose that there are m_k cards of suit k in the pack $(k = 1, 2, \ldots, s)$.

One might then be interested in the *joint* probability that in n draws one receives $R_1, R_2, \ldots, R_s$ cards of suits $1, 2, \ldots, s$ respectively, the actual cards received and their order being immaterial.

Let $I_k(\xi)$ be the indicator variable for the event 'card of kth suit drawn', so that

$$R_k = \sum_{j=1}^{n} I_k(\xi_j) \quad (k = 1, 2, \ldots, s). \qquad \textbf{4.4.1}$$

Consider now a joint p.g.f.

$$\Pi(z_1, z_2, \ldots, z_s) = E\left(\prod_k z_k^{R_k}\right) = m^{-n} \sum_{\xi_1} \sum_{\xi_2} \cdots \sum_{\xi_n} \prod_k z_k^{\sum_j I_k(\xi_j)}$$

$$= \left(\frac{1}{m} \sum_{\xi} z_k^{I_k(\xi)}\right)^n = \left(\frac{\sum m_k z_k}{m}\right)^n = \left(\sum p_k z_k\right)^n, \qquad \textbf{4.4.2}$$

where $\quad p_k = \dfrac{m_k}{m}, \qquad \textbf{4.4.3}$

the fraction of the pack which is of suit k.

If $P_r = P(R_k = r_k; k = 1, 2, \ldots, s)$, r being considered as the vector of the r_ks, then, by definition, P_r is the coefficient of $\prod_k z_k^{r_k}$ in $\Pi(z_1, z_2, \ldots, z_s)$, and so equal to

$$P_r = n! \prod_{k=1}^{s} \frac{p_k^{r_k}}{r_k!} \quad (r_k \geqslant 0, k = 1, 2, \ldots, s; \sum r_k = n). \qquad \textbf{4.4.4}$$

This distribution is known as the *multinomial distribution*, for obvious reasons. It generalizes the binomial distribution, and can be used to describe, for example, the distribution of a poll of n voters among s candidates, or of n molecules among s energy levels, under appropriate basic assumptions.

Exercises

4.4.1   Show that P_r attains its unique maximum near $r_k = np_k$ $(k = 1, 2, \ldots, s)$.

4.4.2   Show from equation **4.4.2** that

$$E\left(\prod_k R_k^{(v_k)}\right) = n^{(\Sigma v_k)} \prod_k p_k^{v_k},$$

and hence that

$$E\left(\frac{R_k}{n}\right) = p_k,$$

$$\mathrm{cov}\left(\frac{R_j}{n}, \frac{R_k}{n}\right) = \frac{p_j(\delta_{jk} - p_k)}{n}.$$

4.4.3 Show that

$$P\left[\left|\frac{R_k}{n_k} - p_k\right| \leqslant \varepsilon_k; k = 1, 2, \ldots, s\right] \geqslant 1 - \frac{1}{n} \sum_1^s \frac{p_k(1 - p_k)}{\varepsilon_k^2}.$$

4.5 Sampling without replacement; the hypergeometric distribution

Consider again the card-drawing problem of the last two sections, but assume now that a card once drawn is *not* replaced in the pack. This is often a more realistic assumption. If one is carrying out a sample survey, for example, and so choosing a sample of n from a population of m, one will normally make sure that a given individual is not included more than once. Or, in the molecule distribution problem mentioned in section 4.3, it may be that no more than one molecule can occupy a given position at once, so that a position once occupied is forbidden to other molecules.

The number of elementary events is now $m(m-1)(m-2)\ldots(m-n+1) = m^{(n)}$, rather than m^n. We shall assume all these to be equally likely, on the grounds of symmetry.

Consider the colour distribution problem of section 4.3. The number of sequences containing r red and $n-r$ black cards in a given order will now be $m_1^{(r)} m_2^{(n-r)}$ rather than $m_1^r m_2^{n-r}$ (since the first red card can be any one of m_1, the second any one of $m_1 - 1$, etc.). The same argument as that leading to relation **4.3.2** now shows that the probability that the sequence contains just r red cards, in arbitrary order, is

$$P_r = \binom{n}{r} \frac{m_1^{(r)} m_2^{(n-r)}}{m^{(n)}} = \frac{\binom{m_1}{r}\binom{m_2}{n-r}}{\binom{m}{n}}. \qquad \textbf{4.5.1}$$

This is the analogue of the binomial distribution, for sampling without replacement, and is known as the *hypergeometric* distribution. If m is large compared with n then the two distributions differ little, but in many cases m and n will be of comparable magnitude.

One interesting application of this model is the survey method known as *capture–recapture sampling*. Consider a geographically well-defined animal population, such as a lake containing m fish. In a first sample of the lake, m_1 fish are caught, marked and returned to the lake. They are given time to mix with their unmarked brethren and recover from this experience, and then a second sample of n is taken. If it is assumed that marked and unmarked fish are equally likely to be caught in the second sample, the distribution of R, the number of marked fish in the second sample, is given by expression **4.5.1**. The problem of interest in practice is how to determine m, the number of fish in the lake, from m_1, n and r, the observed value of R. If one chooses the value of m

that maximizes expression **4.5.1** (a technique of statistical inference known as the *method of maximum likelihood*) then one derives the estimate of the total fish population

$$\hat{m} = \frac{m_1 n}{r}. \tag{4.5.2}$$

This is also the estimate one would obtain by equating the proportions of marked fish in the lake and in the second sample, and so is reasonable.

Because the result of one draw affects the result of the later draws, the direct application of the p.g.f. technique is not so natural in this case (but see exercises 4.5.5 and 5.6.7). However, we can derive the moments of R from the fact that expression **4.5.1** is a probability distribution, so that

$$\sum_r P_r(m, m_1, n) = 1 \tag{4.5.3}$$

is an identity in m, m_1 and n. We have added these three arguments to P_r to emphasize the fact that the distribution does depend upon them, although r is the only argument corresponding to values of a r.v. The summation in expression **4.5.3** is from 0 to $\min(m_1, n)$.

Now

$$E(R^{(v)}) = \sum_r r^{(v)} P_r(m, m_1, n)$$

$$= \sum_r \frac{n^{(v)} m_1^{(v)}}{m^{(v)}} P_{r-v}(m-v, m_1-v, n-v)$$

$$= \frac{n^{(v)} m_1^{(v)}}{m^{(v)}} \quad (v = 0, 1, 2, \ldots). \tag{4.5.4}$$

This corresponds exactly to equation **4.3.4**, with factorial powers replacing ordinary powers.

One can consider the multiple classification problem of section 4.4 for this case also. The probability of observing r_k cards from the kth suit ($k = 1, 2, \ldots, s$) will now be

$$P_r = \frac{n!}{m^{(n)}} \prod_k \frac{m_k^{(r_k)}}{r_k!} = \frac{\prod\limits_{k=1}^{s} \binom{m_k}{r_k}}{\binom{m}{n}}. \tag{4.5.5}$$

Exercises

4.5.1   Show that for the hypergeometric distribution **4.5.1**,

$$E(R) = \frac{nm_1}{m},$$

$$\text{var}(R) = \frac{nm_1}{m}\left(1 - \frac{m_1}{m}\right)\left(\frac{m-n}{m-1}\right).$$

4.5.2   Show that for the vector hypergeometric distribution **4.5.5**,

$$E\left[\prod_k R_k^{(v_k)}\right] = \frac{n^{(\Sigma v_k)}}{m^{(\Sigma v_k)}} \prod_k m_k^{(v_k)}.$$

4.5.3   Modify the estimate **4.5.2** to

$$\hat{m} = \frac{(m_1+1)(n+1)}{r+1} - 1,$$

and show that this has expectation

$$E(\hat{m}) = m - \frac{m_2^{(n+1)}}{m^{(n)}}.$$

4.5.4   Show that for sampling without replacement the distribution of exercise 4.3.2 now becomes

$$P(U = u) = \frac{n(u-1)^{(n-1)}}{m^{(n)}},$$

and that the result

$$E[(U-n)^{(v)}] = \frac{n(m-n)^{(v)}}{n+v} \quad (v = 0, 1, 2, \ldots)$$

holds exactly. Hence

$$E(U) = \frac{n(m+1)}{n+1},$$

$$\mathrm{var}(U) = \frac{n(m+1)(m-n)}{(n+1)^2(n+2)}.$$

4.5.5   Consider the population sampling problem (exercise 4.3.3) in the case when sampling is carried out without replacement. If $x_1, x_2, \ldots, x_n$ are the income values observed in the sample, and $S_n = \sum_1^n x_j$, show that

$$\sum_{n=0}^m \binom{m}{n} \zeta^n E(e^{\theta S_n}) = \prod_{k=1}^m (1 + \zeta e^{\theta y_k})$$

where ζ and θ are constants. Hence, or otherwise, show that, on a sample of n,
$E(\bar{x}) = \bar{y}$,

$$\mathrm{var}(\bar{x}) = \frac{m-n}{mn(m-1)} \sum_1^m (y_k - \bar{y})^2.$$

Compare these with the corresponding results in exercise 4.3.3: note that $\mathrm{var}(\bar{x})$ is zero if $n = m$, as it must be. Show that, if

$$B = \frac{m-n}{mn(n-1)} \sum_1^n (x_j - \bar{x})^2,$$

then $E(B) = \text{var}(\bar{x})$. Thus in practice B could be used as an estimate of $\text{var}(\bar{x})$, and so of the accuracy with which $\bar{x}$ estimates $\bar{y}$.

4.6   Indefinite sampling; the geometric and negative binomial distributions

Let us continue with the card-drawing example, but examine another r.v. Let N_1 be the number of cards one must take (with replacement) before a red card is drawn. With a sample of size n we have

$$P(N_1 = n_1) = \frac{m_2^{n_1-1} m_1 m^{n-n_1}}{m^n} = q^{n_1-1}p, \qquad \textbf{4.6.1}$$

since we must draw $n_1 - 1$ black cards, then a red, and the draw after that is immaterial. Formula **4.6.1** holds only for $n_1 \leqslant n$, for we must reckon with the contingency that a red card is not drawn at all: this event has probability q^n. If $p > 0$ (i.e. if there are any red cards at all in the pack) then one can choose a finite value of n which makes this probability arbitrarily small. The event 'drawing a red card' is thus certain to occur in a finite time and we can introduce the idea of indefinite sampling: that one samples *until* a red card is drawn (implying a changed sample space; see exercise 4.6.6). Under these conditions distribution **4.6.1** holds for the whole range $n_1 = 1, 2, \ldots$.

This distribution is known as the *geometric distribution*. It conceivably holds for many quantities of physical interest: the position of the first fault in a sequence of components, the number of shots needed to strike a target, etc. Unlike previous distributions in this chapter, the range of the variable N_1 is infinite. The probabilities decrease steadily with increasing N_1, rather than coming to a maximum.

One readily finds that

$$\Pi(z) = E(z^{N_1}) = \frac{pz}{1-qz} \quad (|z| < q^{-1}), \qquad \textbf{4.6.2}$$

$$E(N_1) = p^{-1}, \qquad \textbf{4.6.3}$$

$$\text{var}(N_1) = qp^{-2}. \qquad \textbf{4.6.4}$$

Suppose that after the first red card a further N_2 draws are required to reach the second red card, yet another N_3 for the third, and so on. Then, by the same argument that led to **4.6.1** we see that

$$P(N_j = n_j; j = 1, 2, \ldots, r) = \prod_{j=1}^{r} (pq^{n_j-1}). \qquad \textbf{4.6.5}$$

This can only hold for $\sum_{1}^{r} N_j \leqslant n$. However, it is again easy to show that the event 'drawing r red cards' is certain to occur in a finite time, and we can consider a scheme in which one continues sampling until the rth red card is drawn. Note that the probability **4.6.5** factors into separate functions of n_1, of n_2, etc. This is a sign that the r.v.s $N_1, N_2, \ldots, N_r$ are *statistically independent* (see section 5.5).

The number of draws needed to reach the rth red card is a r.v.

$$N = \sum_{j=1}^{r} N_j. \qquad \textbf{4.6.6}$$

Let us consider the distribution of this quantity. By formula **4.6.5** we have

$$E(z^N) = E(z^{\Sigma N_j}) = \left(\frac{pz}{1-qz}\right)^r. \qquad \textbf{4.6.7}$$

Picking out the coefficient of z^n in this expression, we see that

$$P(N = n) = \binom{-r}{n-r} p^r(-q)^{n-r}$$

$$= \binom{n-1}{r-1} p^r q^{n-r} \quad (n = r, r+1, \ldots). \qquad \textbf{4.6.8}$$

This is the *negative binomial distribution*, which provides an interesting contrast to the positive binomial distribution **4.3.2**. The variables n and r have the same connotations in the two cases: the size of the sample and the number of red cards drawn, respectively. However, in case **4.3.2** n is fixed and r is the value of a r.v.; in case **4.6.8** the roles are reversed. This second type of sampling is known as *inverse sampling*; it is often advantageous if p is rather small. Thus, suppose one wishes to estimate the frequency p of a rare blood group, for which p is small. Rather than testing a fixed number of subjects, one gains by sampling until a certain number of the rare group have been observed.

For $r > 1$ expression **4.6.8** does in fact have a maximum, near $n = (r-1)/p$. The p.g.f. of N is of course given by formula **4.6.7**, and

$$E(N) = \frac{r}{p}, \qquad \textbf{4.6.9}$$

$$\text{var}(N) = \frac{rq}{p^2}. \qquad \textbf{4.6.10}$$

There are many other 'first occurrence' problems; see, for example, exercise 4.6.5.

Exercises

4.6.1   Show that $E[(N_1-1)^{(v)}] = v! \, (q/p)^v$.

4.6.2 Show that, if sampling is without replacement, then

$$P(N_1 = n) = \frac{m_2^{(n-1)}m_1}{m^{(n)}} \quad (n = 1, 2, \ldots, m_2 + 1),$$

$$E[(m_2 - N_1 + 1)^{(v)}] = \frac{m_2^{(v)}m_1}{m_1 + v} \quad (v = 0, 1, 2, \ldots).$$

Note that $m_2 - N_1 + 1$ is the number of black cards remaining in the pack when the first red one is drawn.

4.6.3 Give a direct combinatorial derivation of the negative binomial distribution **4.6.8**.

4.6.4 Show that for the negative binomial distribution **4.6.8**

$$E[(N - r)^{(v)}] = (r + v - 1)^{(v)}\left(\frac{q}{p}\right)^v,$$

$$E[(N + v - 1)^{(v)}] = \frac{(r + v - 1)^{(v)}}{p^v} \quad (v = 0, 1, 2, \ldots).$$

4.6.5 Suppose that one samples from the m cards, with replacement, until each of the cards has occurred at least once. Show that, if N is the size of this random sample, then

$$E(z^N) = \prod_{j=1}^{m} \left(\frac{p_j z}{1 - q_j z}\right),$$

where $\quad p_j = \frac{m - j + 1}{m},$

and hence that

$$E(N) = m \sum_{1}^{m} j^{-1},$$

which is asymptotic to $m \log m$ for large m.

The interest of this example is that it is just the 'picture card' problem. Suppose a packet of cereal contains one picture card from a set of m: N is the number of packets one must buy in order to complete the set. Note that $E(N)$ increases faster than m: big sets are relatively more difficult to complete than small ones.

4.6.6 By allowing indefinite sampling we have now moved from the sample space of sequences of length n, with m^n points, to that of all sequences, with an infinite number of points. Let ξ_j denote the outcome of the jth trial, and let $X_j(\xi_j)$ be a r.v. defined on this outcome. Show that we are in effect assuming the expectation functional

$$E\left[\prod_1^\infty X_j\right] = \prod_1^\infty E(X_j) = \prod_1^\infty \left[\frac{\sum\limits_\xi X_j(\xi)}{m}\right].$$

4.7 Sampling from a continuum; the Poisson process

A final variant on our problem will be to assume that ξ, instead of taking m equally likely values, has a uniform distribution over a continuous set T, so that expression **4.3.1** for the expectation of an arbitrary r.v. now becomes

$$E(X) = V^{-n} \int_T d\xi_1 \int_T d\xi_2 \ldots \int_T d\xi_n \ X(\xi_1, \xi_2, \ldots, \xi_n), \qquad \textbf{4.7.1}$$

where V is the content

$$V = \int_T d\xi \qquad \textbf{4.7.2}$$

of T. In the simplest case T is an interval on the real line, of length V. One might be imagining that n molecules are distributed over this interval, all distributions being equally likely, in the sense **4.7.1**. In effect, one is assuming that a molecule is uniformly distributed over T, i.e. is equally likely to lie in any ξ-interval of a given length, and that there is no interaction between molecules. There is no distinction in this case between sampling with or without 'replacement', because integral **4.7.1** will not be affected if $\xi_j = \xi_k$ is forbidden for $j \neq k$. Somewhat more generally, T could be a region in a Euclidean space of higher dimension than one, of 'volume' V, with vector coordinate ξ, and $d\xi$ the infinitesimal element of volume. For example, one might be considering the distribution of molecules in physical space of three dimensions.

Now, consider a region T_1 in T, of content V_1. Just as in section 4.3, it follows that the distribution of the number R of molecules in T_1 is given by the binomial expression **4.3.2**, with

$$p = \frac{V_1}{V}. \qquad \textbf{4.7.3}$$

Suppose however that we let T and n increase indefinitely in such a way that

$$\rho = \frac{n}{V} \qquad \textbf{4.7.4}$$

remains constant. That is, we conceive that T_1 (held fixed in description and size) is part of an indefinitely large region in space, T, but that the number of molecules is also increased so as to hold the average molecule density ρ constant. The process obtained in this way is called a *Poisson process* of intensity ρ.

Taking this limit in expression **4.3.2** for a fixed n we find that

$$P(R = r) = \lim_{n \to \infty} \binom{n}{r} \left(\frac{\rho V_1}{n}\right)^r \left(1 - \frac{\rho V_1}{n}\right)^{n-r}$$

$$= \frac{(\rho V_1)^r}{r!} \lim \left(1 - \frac{\rho V_1}{n}\right)^{n-r} \frac{n(n-1)\ldots(n-r+1)}{n^r}$$

$$= \frac{e^{-\rho V_1}(\rho V_1)^r}{r!}. \tag{4.7.5}$$

The distribution

$$P_r = \frac{e^{-\lambda}\lambda^r}{r!} \quad (r = 0, 1, 2, \ldots) \tag{4.7.6}$$

is known as the *Poisson distribution with parameter* λ. We find readily that on the basis of distribution **4.7.6**

$$\Pi(z) = E(z^R) = e^{\lambda(z-1)}, \tag{4.7.7}$$

$$E(R^{(\nu)}) = \lambda^\nu, \tag{4.7.8}$$

$$E(R) = \text{var}(R) = \lambda. \tag{4.7.9}$$

We conclude then that, if T_1 can be regarded as part of an infinite region in which molecules are distributed at random with average density ρ, then R is distributed as a Poisson variable with expected value ρV_1. The difference between the binomial and Poisson distributions is that the constraining effect of the finiteness of n and V is removed in the second case.

The Poisson distribution is sometimes known as the 'law of rare events': it tends to occur when one has a large number of trials, each with only a small chance of a positive result. So, in the present case, the chance that any *given* molecule will lie in T_1 is small, of order V^{-1}. However, since n is of the same order as V, the expected number of molecules in T_1 is ρV_1, and so always appreciable.

We could have applied the limiting procedure $n, V \to \infty$ to the p.g.f. of R rather than to the individual probability: this procedure is in fact justifiable, and rather more elegant. Let us apply it to calculate the joint probability distribution of the numbers of molecules $R_1, R_2, \ldots, R_s$ in non-overlapping regions $T_1, T_2, \ldots, T_s$. We shall assume that T_k has content V_k, and shall set

$$p_k = \frac{V_k}{V}. \tag{4.7.10}$$

In general $\sum p_k < 1$, because we shall not assume that $\bigcup T_k = T$.

Then, by the same derivation as that of equation **4.4.2**, we find that

$$\Pi(z) = E\left(\prod_k z_k^{R_k}\right) = (1 - \sum p_k + \sum p_k z_k)^n. \tag{4.7.11}$$

Let us now apply the same limiting procedure as before, holding the regions T_k fixed. We find then that

$$\Pi(z) = \left[1 + \frac{\rho}{n}\sum V_k(z_k - 1)\right]^n \to \exp[\rho \sum V_k(z_k - 1)],\qquad \textbf{4.7.12}$$

so that

$$P(R_k = r_k; k = 1, 2, \ldots, s) = \prod_{k=1}^{s} \frac{e^{-\rho V_k}(\rho V_k)^{r_k}}{r_k!}.\qquad \textbf{4.7.13}$$

The fact that this probability factorizes into the individual probabilities for the R_k indicates that the R_k are independently distributed, in the sense to be explained in section 5.5.

Exercises

4.7.1   Show that the Poisson distribution **4.7.6** has a peak near λ if $\lambda > 1$, but is otherwise monotone decreasing.

4.7.2   Show that if R follows distribution **4.7.5** then $E[(R/V_1 - \rho)^2] = \rho/V_1$, so that $R/V_1 \xrightarrow{\text{m.s.}} \rho$ as $V_1 \to \infty$.

4.7.3   Suppose that regions T_1 and T_2 have contents V_1 and V_2 respectively, and overlap by an amount V_0. Show that for the Poisson process

$$E(z_1^{R_1} z_2^{R_2}) = \exp[\rho V_1(z_1 - 1) + \rho V_2(z_2 - 1) + \rho V_0(z_1 - 1)(z_2 - 1)].$$

4.8   'Nearest neighbours'; the exponential and gamma distributions

Consider the Poisson process of the previous section. We see from equation **4.7.5** that the probability that a region of content V_1 contains no molecules is just $e^{-\rho V_1}$. Suppose that T is the real axis, and T_1 the interval $(x_0, x_0 + x)$. The probability that there is no molecule within a distance x to the right of a point x_0 is thus $e^{-\rho x}$. Let X be the r.v. 'distance from x_0 to the next molecule on the right'. Our probability is then just

$$P(X > x) = e^{-\rho x}.\qquad \textbf{4.8.1}$$

The fact that expression **4.8.1** is differentiable implies the existence of a probability density for X (see exercise 2.4.4):

$$f(x) = \frac{\partial}{\partial x}P(X \leqslant x) = \rho e^{-\rho x} \quad (x \geqslant 0).\qquad \textbf{4.8.2}$$

This is the density of the *exponential* distribution, the continuous equivalent of the geometric distribution **4.6.1**. In fact, the r.v. is of the same character in the two cases: distance or time or number of trials needed for a 'first occurrence' of some event. The exponential distribution is used very extensively, for study of quantities such as failure times, lifetimes, distance between defects, etc.

The calculation would not have been affected if we had taken x_0 as the co-ordinate of a particular molecule, since the knowledge that one molecule is at x_0 does not affect the distribution of the others. Hence **4.8.3** is also the density function of the distance between consecutive molecules.

Just as consideration of rth occurrences leads us to generalize the geometric distribution **4.6.1** to the negative binomial distribution **4.6.8**, so can the exponential distribution be generalized. Let the rth molecule with coordinate greater than x_0 have coordinate $x_0 + X$, and let the number of molecules in the interval $(x_0, x_0 + x)$ be denoted by R_x. Then the events $X > x$ and $R_x < r$ are identical, so that

$$P(X > x) = P(R_x < r)$$

$$= \sum_{j=0}^{r-1} \frac{e^{-\rho x}(\rho x)^j}{j!}. \qquad 4.8.3$$

Differentiating as in **4.8.2** we then obtain the probability density of X as

$$f(x) = \frac{\rho^r e^{-\rho x} x^{r-1}}{(r-1)!} \quad (x \geqslant 0). \qquad 4.8.4$$

This is termed the *gamma distribution*, and is the continuous equivalent of the negative binomial distribution **4.6.8**.

Exercises

4.8.1   Show that for the gamma distribution **4.8.4**

$$E(X^v) = \frac{(r+v-1)^{(v)}}{\rho^v} \quad (v = 0, 1, 2, \ldots).$$

4.8.2   Consider a Poisson process in a Euclidean space of a dimensions. Let the rth nearest molecule to a chosen origin be distant X_r from the origin, and let a hypersphere of radius X_r have content A_r ($A_0 = 0$). Show that

$$E(A_r - A_{r-1}) = \frac{1}{\rho} \quad (r = 1, 2, \ldots).$$

What is the probability density of X_1?

4.9   Other simple processes

There is a whole host of allocation and occupation problems which can be reduced to the study of a simple process: one with equiprobable elementary events. We have studied the commonest and most useful ones in the foregoing sections, but there are many others; perhaps more of interest in themselves, or as a challenge to one's combinatorial ingenuity, than of very general significance.

4.9.1 *Occupancy statistics*

We have considered the situations where one samples n times from m quantities with and without replacement. The numbers of elementary events are respectively m and $m^{(n)}$ in these two cases, and for the situation as we have described it, one can reasonably assume these elementary events to be equally probable.

However, in some physical applications one is forced to another specification, in which the *order* of events is regarded as intrinsically unobservable and meaningless. That is, for example, that different orders of drawing a given set of cards cannot be distinguished, and the elementary event is simply that one has drawn n_k cards of type k ($k = 1, 2, \ldots, m$), order unspecified. These events are regarded as equally likely. In our previous descriptions such an event would have been regarded as made up of

$$\frac{n!}{\prod_1^m n_k!} \qquad\qquad\qquad\qquad \textbf{4.9.1}$$

elementary events, and its probability would have been correspondingly weighted (although in the 'without replacement' case all the n_k would be 0 or 1). Now this factor is absent.

If K_{mn} is the number of elementary events for this case then

$$K_{1n} = 1, \qquad\qquad\qquad\qquad \textbf{4.9.2}$$

$$K_{mn} = \sum_{r=0}^{n} K_{m-1,r}. \qquad\qquad\qquad\qquad \textbf{4.9.3}$$

The $K_{m-1,r}$ in sum **4.9.3** is the number of events obtained by sampling r times from a pack of $m-1$ cards, and then sampling the mth card $n-r$ times – the order in which these are to be intermingled no longer matters. From **4.9.2** and **4.9.3** we find that the number of elementary events is

$$K_{mn} = \frac{(m+n-1)^{(n)}}{n!}. \qquad\qquad\qquad\qquad \textbf{4.9.4}$$

All three types of statistics are important in statistical physics: we shall summarize the situation in a table. Since it is the physical applications which are important, we shall speak of n molecules distributed among m energy levels rather than of n draws from a pack of m cards.

Type of statistics	Determining properties	Number of elementary events	Probability of occupation numbers n_k	Generating function
Boltzmann	The molecules are distinguishable, and sampling is 'with replacement' – multiple occupation of an energy level is permitted	m^n	$\dfrac{n!}{m^n \prod_1^m n_k!}$	$\exp\left(\sum_1^m z_k\right)$
Fermi–Dirac	'Sampling without replacement': a given energy level can be occupied at most once	$m^{(n)}$	$\dfrac{n!}{m^{(n)}}$	$\prod_1^m (1+z_k)$
Bose–Einstein	'Order of sampling' unobservable and meaningless: the n molecules are indistinguishable	$\dfrac{(m+n-1)^{(n)}}{n!}$	$\dfrac{n!}{(m+n-1)^{(n)}}$	$\prod_1^m (1-z_k)^{-1}$

The 'occupation numbers' n_k of the fourth column are the number of times each particular card is drawn, or energy level occupied ($k = 1, 2, \ldots, m$). The corresponding probabilities are generated by the function in the last column, in that, if this function is denoted by $\phi(z_1, z_2, \ldots, z_m)$ then

$P(\text{occupation numbers } n_k; k = 1, 2, \ldots, m)$

$$= \frac{\text{coefficient of } \prod_1^m z_k^{n_k} \text{ in } \phi(z_1, z_2, \ldots, z_m)}{\text{coefficient of } z^n \text{ in } \phi(z, z, \ldots, z)}. \qquad \textbf{4.9.5}$$

Large values of individual n_k are relatively more probable on Bose–Einstein than on Boltzmann statistics; they are, of course, prohibited on Fermi–Dirac statistics.

4.9.2   Matching problems

There is a problem appearing under many guises, one of which is this: n letters are written to different people, and envelopes correspondingly addressed. The letters are now mixed before being put in the envelopes. What is the probability A_j that just j letters are placed in their correct envelopes? One imagines that the effect of the mixing is to make all $n!$ possible arrangements equally likely. For brevity, we shall say that a letter 'matches' if it is in the correct envelope.

Suppose P_j is the probability that a given j letters match and none of the others do. Since

$$A_j = \binom{n}{j} P_j, \qquad\qquad \textbf{4.9.6}$$

the problem is solved if the P_j can be calculated – but this is by no means immediate. The quantities which can be immediately calculated are:

$Q_j = P($a given j letters match; no condition imposed on the remaining $n-j$)

$$= \frac{(n-j)!}{n!} = \frac{1}{n^{(j)}}. \qquad\qquad \textbf{4.9.7}$$

We shall leave it to the reader to verify the relation

$$Q_j = \sum_{k=j}^{n} \binom{n-j}{k-j} P_k; \qquad\qquad \textbf{4.9.8}$$

and to show, by means of generating functions or otherwise, that equation **4.9.8** can be inverted to

$$P_j = \sum_{k=j}^{n} \binom{n-j}{k-j}(-1)^{k-j} Q_k. \qquad\qquad \textbf{4.9.9}$$

From **4.9.6, 4.9.7** and **4.9.9** we find

$$A_j = \frac{1}{j!} \sum_{k=j}^{n} \frac{(-1)^{k-j}}{(k-j)!}. \qquad\qquad \textbf{4.9.10}$$

There is another method which is much faster, although perhaps less obvious. Denote A_j by A_{nj} instead, to emphasize its dependence upon n, and define the p.g.f.

$$\Pi_n(z) = \sum_{j=0}^{n} A_{nj} z^j. \qquad\qquad \textbf{4.9.11}$$

Let $B_{nj} = n! A_{nj}$, the number of the $n!$ arrangements that yield exactly j matches. Then

$$B_{nj} = \frac{j+1}{n+1} B_{n+1,j+1}, \qquad\qquad \textbf{4.9.12}$$

so that

$$A_{nj} = (j+1)A_{n+1,j+1}. \qquad\qquad \textbf{4.9.13}$$

Relation **4.9.12** follows from the fact that we can view the j-match case for n letters as a $(j+1)$-match case for $n+1$ letters, with the $(n+1)$th letter always among the matched ones. But, by symmetry, in only a fraction $(j+1)/(n+1)$

of the $B_{n+1,j+1}$ cases will the $(n+1)$th letter fall among the $j+1$ which are matched.

Relation **4.9.13** is equivalent to

$$\frac{\partial \Pi_{n+1}}{\partial z} = \Pi_n,$$ **4.9.14**

or $\quad \Pi_{n+1}(z) = 1 + \int_1^z \Pi_n(w)\, dw,$ **4.9.15**

since $\Pi_n(1) = 1$ for all n. Applying relation **4.9.15** repeatedly, starting from $\Pi_1 = z$, we obtain

$$\Pi_n(z) = \sum_{k=0}^n \frac{(z-1)^k}{k!},$$ **4.9.16**

a truncated version of the Poisson p.g.f. The coefficient of z^j in this expression is indeed equal to expression **4.9.10**.

4.9.3   *Nearest neighbours and lattice statistics (the Ising problem)*

We end with an unresolved question, to show how easily a simple process can generate a deep combinatorial problem. Suppose that n_A molecules of type A and n_B molecules of type B are ordered randomly along a line, in such a way that all $(n_A + n_B)!$ possible orderings are equally likely. If n_{AB} is the number of times an A molecule is followed in the sequence by a B molecule, etc., one asks for the joint distribution of n_{AA}, n_{AB}, n_{BA} and n_{BB}. The point of this calculation is that, if only 'nearest neighbours' interact, an AB pair having an energy of interaction H_{AB}, etc., then the total energy of interaction is

$$n_{AA}H_{AA} + n_{AB}H_{AB} + n_{BA}H_{BA} + n_{BB}H_{BB},$$

and physical arguments show (see section 6.5) that it is this that determines which configurations actually predominate.

This problem is soluble by elementary methods: the reader may care to attempt it. The corresponding problem for molecules on a square lattice in two dimensions has been partially solved, with great difficulty. For three or more dimensions the problem is unsolved, despite its importance for theories of ferromagnetism and crystal structure.

Exercises

4.9.1   Let us consider the situation and nomenclature of section 4.4, but suppose that Bose–Einstein statistics prevail. Show that, instead of the multinomial expression **4.4.4**, we have

$$P_r = \frac{n!}{(m+n-1)^{(n)}} \prod_{k=1}^s \frac{(m_k + r_k - 1)^{(r_k)}}{r_k!}.$$

(Hint: show from expression **4.9.5** that this probability is proportional to the coefficient of $\Pi z_k^{r_k}$ in $\Pi(1-z_k)^{-m_k}$.)

4.9.2   Consider the matching problem of sub-section (b). The following argument is incorrect; locate the fallacy: 'The probability of at least j matches is $\binom{n}{j} Q_j$; hence

$$A_j = \binom{n}{j} Q_j - \binom{n}{j+1} Q_{j+1} = \frac{j}{(j+1)!},$$

4.9.3   If J is the number of matches, show that

$$E(J^{(\nu)}) = \begin{cases} 1 & (\nu \leqslant n), \\ 0 & (\nu > n). \end{cases}$$

4.9.4   Show that one can express A_0 as

$$A_0 = \frac{[n!/e]}{n!},$$

where $[x]$ is the integer nearest to x.

79   **Other simple processes**

5 Conditioning

5.1 Conditional expectation

The idea behind *conditioning* is that a partial observation can give one some idea of where in sample space the ω corresponding to the full observation must lie: in a set A, say. Then, in effect, the sample space has contracted: instead of taking expectations over the full space Ω, one does so only over A.

For example, suppose a doctor is to examine a patient, and that ω represents 'the patient's physical condition' as revealed by a standard medical examination. We suppose the doctor has no foreknowledge of the patient, so that, if Ω is the sample space representing the whole range of condition he encounters in his practice, then expectations must be taken over Ω before examination.

However, suppose that as soon as the patient enters the room the doctor notices that he suffers from shortness of breath, with its implications of damage to lungs or heart. In effect, the doctor has performed a partial observation, and the patient is no longer 'any patient' (with ω in Ω) but 'a patient with shortness of breath' (with ω in A, if A corresponds to the event 'suffers from shortness of breath'). The doctor's expectations will now be changed, because he is dealing with a more specific case. For example, his expectation of the r.v. 'extent of lung damage' will certainly have increased, because patients in set A are known to have greater lung damage on the average than patients in general.

So, the doctor would carry out further tests, narrowing down the potential sample space every time, until at last a confident diagnosis is possible, and ω has been fully determined for the particular patient.

We shall now define 'expectation of X conditional on A' as

$$E(X|A) = \frac{E(XI_A)}{E(I_A)} = \frac{E(XI_A)}{P(A)},$$

5.1.1

where $I_A(\omega)$ is, as usual, the indicator function of the set A. Definition **5.1.1** is reasonable. The inclusion of I_A in the numerator means that $X(\omega)$ is averaged only for A, the more specific set to which ω is known to belong. However, in contracting the sample space we must 'renormalize' the expectation, in order to retain the property of an average, $E(1|A) = 1$. We therefore divide by $P(A)$.

Operation **5.1.1** corresponds exactly to what one does when one carries out a 'partial average', in the population census of section 1.3, for example. Thus, suppose $X(\omega)$ is 'income', then average income in the community is

$$\frac{\sum n_k X(\omega_k)}{\sum n_k}, \qquad\qquad \textbf{5.1.2}$$

where k runs over the census categories. However, suppose we are interested in 'average income for males' so that the conditioning event A is 'male sex'. Then the sums in numerator and denominator of expression **5.1.2** would be restricted to male categories, which is exactly what we are doing in **5.1.1**.

This example makes it clear that the conditioning event must be one which is observable on Ω: one must be able to say that each ω either does or does not belong to A. For the census example one could not calculate 'average male income' if sex had not been recorded in the census, because the number of males in each category would then be indeterminate.

Also it is necessary that there should be *some* males, i.e. $\sum\limits_{\text{male}} n_k > 0$, otherwise the partial average is meaningless, and, indeed, takes the indeterminate form 0/0. The corresponding condition in operation **5.1.1** is $P(A) > 0$, and we shall retain this condition for the next section or two. However, one does in fact run quite quickly into situations where it is not only reasonable, but also necessary, to consider conditioning events of zero probability. These cases require some caution, and will be considered especially in section 5.3.

Let us consider some examples. One conditional expectation that interests us all is 'expectation of life'. Suppose the length of a human life can be regarded as a r.v. X with density function $f(x)$. A person has survived to age a: how much longer can he expect to live? The conditioning event A is that the person has survived to age a, equivalent to '$X > a$'. The time he has yet to live is $X - a$, and the conditional expectation of this quantity is the 'expected residual life at age a':

$$E(X - a \mid X > a) = \frac{\int\limits_{a}^{\infty} (x - a) f(x)\, dx}{\int\limits_{a}^{\infty} f(x)\, dx}, \qquad\qquad \textbf{5.1.3}$$

by definition **5.1.1**.

Suppose that R_1 and R_2 are two r.v.s with joint p.g.f. $\Pi(z_1, z_2) = E(z_1^{R_1} z_2^{R_2})$. Then one sees from **5.1.1**, with a little thought, that the p.g.f. of R_1 conditional on a value of R_2 is

$$E(z^{R_1} \mid R_2 = r_2) = \frac{\text{coefficient of } z_2^{r_2} \text{ in } \Pi(z, z_2)}{\text{coefficient of } z_2^{r_2} \text{ in } \Pi(1, z_2)}. \qquad\qquad \textbf{5.1.4}$$

Consider the case **4.7.12** for the p.g.f. of numbers of molecules in disjoint regions in a Poisson process, when

$$\Pi(z_1, z_2) = \exp[\rho V_1(z_1 - 1) + \rho V_2(z_2 - 1)].\qquad \textbf{5.1.5}$$

Because this factorizes into individual p.g.f.s we see from equation **5.1.4** that

$$E(z^{R_1} | R_2 = r_2) = \exp[\rho V_1(z - 1)],$$

which is functionally independent of r_2. Thus R_1 has the same Poisson distribution, whatever the value of R_2. This reflects the *statistical independence* of R_1 and R_2 (see section 5.5).

Suppose we consider the p.g.f. of R_1 conditional on $R_1 + R_2 = r$, that is, on the assumption that the total number of molecules $R = R_1 + R_2$ in the two regions is known, and equal to r. Since R_1 and R have joint p.g.f.

$$E(z_1^{R_1} w^R) = \Pi(z_1 w, w),$$

where Π is given by equation **5.1.5**, we see from **5.1.4** that

$$E(z^{R_1} | R_1 + R_2 = r) = \frac{\text{coefficient of } w^r \text{ in } \exp[\rho V_1 zw + \rho V_2 w]}{\text{coefficient of } w^r \text{ in } \exp[\rho V_1 w + \rho V_2 w]}$$

$$= \left(\frac{V_1 z + V_2}{V_1 + V_2}\right)^r = (pz + q)^r,$$

where $p = V_1/(V_1 + V_2)$. That is, the conditional distribution of R_1 is binomial; not a surprising conclusion when one recalls how the Poisson process was constructed.

The principal formal properties of a conditional expectation are summed up in the following theorem.

Theorem 5.1.1

(i) *If $P(A) > 0$ then the conditional expectation* **5.1.1** *obeys the axioms for an expectation of section* 2.2.

(ii) *If $\{A_k\}$ is a decomposition of Ω, then*

$$E(X) = \sum_k P(A_k) E(X | A_k).\qquad \textbf{5.1.6}$$

The proof of the first part is direct. The second assertion follows from the fact that the right-hand member of relation **5.1.6** is just

$$\sum E(X I_{A_k}) = E(X \sum I_{A_k}) = E(X),$$

by exercise 3.2.1.

Relation **5.1.6** is basic, and can be regarded as at least a partial inverse to **5.1.1**. Whereas **5.1.1** gives the rule for the transformation of E when a sample space is *contracted* from Ω to A; relation **5.1.6** gives the transformation rule when a sample space is expanded from individual spaces A_k to an over-all space Ω, by averaging with respect to an extra variable. It can be regarded as a formula which constructs an expectation in two stages; first by averaging *within* an event A_k, and then by averaging *over* events $A_1, A_2, \ldots$. This is an aspect that will recur.

For example, suppose electric lamps may be of various manufactures, and that lamps of brand k have a lifetime X with probability density

$$f_k(x) = \lambda_k e^{-\lambda_k x}.\qquad\qquad\textbf{5.1.7}$$

Lamps of brand k thus have expected lifetime

$$E(X\,|\,A_k) = \int_0^\infty x f_k(x)\,dx = \lambda_k^{-1}.\qquad\qquad\textbf{5.1.8}$$

We have written this as $E(X\,|\,A_k)$ because we are going to regard 'the lamp is of manufacture k' as an event A_k, and relation **5.1.8** is then an expectation conditional on this event.

Suppose now that one buys a lamp at random, and that this has probability $P(A_k) = \pi_k$ of being of manufacture k. Its expected lifetime is then, by relations **5.1.6** and **5.1.8**,

$$E(X) = \sum \pi_k \lambda_k^{-1}.\qquad\qquad\textbf{5.1.9}$$

This expectation has been obtained in two stages: by averaging in relation **5.1.8** over the fluctuations in lifetime of a lamp of given manufacture, and then by averaging in **5.1.9** over k, the different sources of manufacture.

There are various special cases of a conditional expectation. For example, the probability of an event B, conditional on the occurrence of A, is

$$P(B\,|\,A) = E(I_B\,|\,A) = \frac{E(I_A I_B)}{E(I_A)} = \frac{P(AB)}{P(A)}.\qquad\qquad\textbf{5.1.10}$$

If there exists a function $f(x\,|\,A)$ such that

$$E[H(X)\,|\,A] = \int H(x) f(x\,|\,A)\,dx$$

for a class of functions H which includes at least the step functions, then $f(x\,|\,A)$ will be termed the *probability density function of X conditional on A*. Thus expression **5.1.7** could be written $f(x\,|\,A_k)$.

Exercises

5.1.1   Show for the example of exercise 4.7.3 that

$$E(R_2\,|\,R_1 = r_1) = \rho V_2 + \frac{V_0}{V_1}(r_1 - \rho V_1)$$

and interpret. Calculate the conditional variance of R_2.

5.1.2   The r.v.s. $R_1, R_2, \ldots, R_s$ follow the multinomial distribution **4.4.4**. Show that

$$E\left(\prod_1^t z_k^{R_k}\,\middle|\,R_j = r_j; j \geqslant t\right) = \left(\frac{\sum_1^t p_k z_k}{\sum_1^t p_k}\right)^{n - \sum_{t+1}^s r_j}$$

and hence that the conditioned distribution of $R_1, R_2, \ldots, R_t$ is also multi-nomial.

5.1.3   For the example associated with relation **5.1.3**, what is the conditional probability density of residual lifetime?

5.1.4   Suppose the probability that a lamp fails in the age interval $(x, x+dx)$, conditional on the fact that it has survived to age x, is $\lambda(x)\,dx + o\,(dx)$. That is, $\lambda(x)$ is the *age-specific failure rate* (in the sense of a rate per unit time). Show that the probability that the lamp survives to age x at least is

$$\exp\left[-\int_0^x \lambda(u)\,du \right].$$

Thus, an exponential life distribution corresponds to a constant failure rate.

5.1.5   Suppose that lamps manufactured by factory k have constant failure rate λ_k, and that π_k is the probability that a randomly chosen lamp is from factory k. Show that for a randomly chosen lamp

$$\lambda(x) = \frac{\sum \pi_k \lambda_k e^{-\lambda_k x}}{\sum \pi_k e^{-\lambda_k x}},$$

and that the expected residual life at age x is

$$L(x) = \frac{\sum \pi_k \lambda_k^{-1} e^{-\lambda_k x}}{\sum \pi_k e^{-\lambda_k x}}.$$

Show also that these quantities are respectively decreasing and increasing functions of x. The terms with small λ_k predominate in these sums as x becomes large. This is 'survival of the fittest' (by attrition rather than competition): those lamps which have survived long are most probably of good manufacture.

5.2   Conditional probability

The conditional probability **5.1.10** is analogous to a partial proportion, just as a conditional average is analogous to a partial average. So, to return to the census example of section 1.3, if A is the event 'male' and B the event 'employed', then $P(B)$ is analogous to 'the proportion of the population which is employed', but $P(B|A)$ is analogous to 'the proportion of the *male* population which is employed'.

The processes of chapter 4 abound in aspects which are usefully viewed conditionally. For example, suppose A is the event 'the first n draws from a pack of m_1 red and m_2 black cards $(m_1 + m_2 = m)$ are all red' and B the event 'the $(m+1)$th draw is red'. If sampling is without replacement then

$$P(A) = \frac{m_1^{(n)}}{m^{(n)}}, \quad \text{and} \quad P(AB) = \frac{m_1^{(n+1)}}{m^{(n+1)}},$$

so that

$$P(B|A) = \frac{m_1 - n}{m - n}.$$

On the other hand, if sampling is with replacement, then

$$P(B|A) = \left(\frac{m_1}{m}\right)^{n+1}\left(\frac{m_1}{m}\right)^{-n} = \frac{m_1}{m}.$$

In the second case the probability is constant, and just equal to the proportion of red cards in the pack; in the first case it is lowered by the fact that the proportion of red cards left in the pack has been reduced by earlier draws.

Exercises

5.2.1   Show that

$$P(ABC) = P(A)P(B|A)P(C|AB),$$

and generalize this result.

5.2.2   A family with two children can have any of the four constitutions bb, bg, gb, gg, where b = boy, g = girl, and account is taken of order. Suppose these four possibilities are equally probable. Show that

$P(\text{two boys}|\text{elder child a boy}) = \frac{1}{2},$

$P(\text{two boys}|\text{at least one boy}) = \frac{1}{3}.$

One is inclined to think these probabilities should be equal, since in both cases one is given the information that the family contains at least one boy, and the extra information given in the first case (that this boy can be labelled as the elder) seems irrelevant. However, A contains two elementary events in the first case, and three in the second. Note that in both cases $AB = B$, since $B \subset A$.

5.2.3   Cards are dealt to four players (all 52! orders of dealing being equally likely). Show that

$$P(\text{the first player holds all aces}|\text{he holds the ace of hearts}) = \binom{48}{9}\bigg/\binom{51}{12}$$

$$= \frac{132}{12\,495},$$

$P(\text{the first player holds all aces}|\text{he holds at least one ace})$

$$= \binom{48}{9}\bigg/\left[\binom{52}{13} - \binom{48}{9}\right] = \frac{33}{12\,462}.$$

Note that we can rewrite formula **5.1.10** as

$$P(AB) = P(A)P(B|A), \qquad\qquad\qquad\qquad \textbf{5.2.1}$$

which can be read: 'The probability that A and B both occur equals the probability that A occurs times the probability of B conditional on the occurrence of A.' This formula is valid even if $P(A) = 0$. From the symmetry of $P(AB)$ with respect to A and B we see that formula **5.2.1** implies

$$P(A|B) = \frac{P(A)P(B|A)}{P(B)}. \qquad \textbf{5.2.2}$$

A particular case of relation **5.1.6** is

$$P(B) = \sum_k P(A_k)P(B|A_k), \qquad \textbf{5.2.3}$$

where $\{A_k\}$ is a decomposition of Ω. This is sometimes known as the 'generalized addition law of probability', and in many cases provides the only way to calculate $P(B)$, in that it is the quantities in the right-hand member of **5.2.3** which are given. For instance, take the lamp example with which we concluded the previous section, and consider the event B, that 'the randomly chosen lamp survives at least to age x'. By **5.2.3** this is

$$P(B) = \sum \pi_k \int_x^\infty f_k(u)\,du = \sum \pi_k e^{-\lambda_k x}. \qquad \textbf{5.2.4}$$

Combining **5.2.2** and **5.2.3**, we see that

$$P(A_j|B) = \frac{P(A_j)P(B|A_j)}{\sum_k P(A_k)P(B|A_k)}. \qquad \textbf{5.2.5}$$

This result is known as *Bayes's theorem*. Although mathematically trivial, it has become celebrated for its quasi-philosophical implications. We shall discuss these in a moment, but first we consider some of its straightforward applications.

Returning to the lamp example, equation **5.2.5** yields

$$P(A_j|B) = \frac{\pi_j e^{-\lambda_j x}}{\sum \pi_k e^{-\lambda_k x}} \qquad \textbf{5.2.6}$$

as the probability that the lamp is of manufacture j, given that it has not yet failed at age x. As mentioned in exercise 5.1.6, as x increases, expression **5.2.6** will become largest for those j for which λ_j, the failure rate, is small: a lamp that has lasted well is probably one of good manufacture.

The quantities $P(A_j)$ and $P(A_j|B)$ are respectively known as the *prior* and *posterior* probabilities of A_j; prior and posterior, that is, to the observation of the event B.

One can equally well find use for Bayes's formula in the medical diagnosis problem with which we began section 5.1. Suppose a patient may be suffering from various conditions, which we shall label A_k $(k = 1, 2, \ldots)$. For simplicity, we shall assume these to be mutually exclusive and exhaustive, so that $\{A_k\}$ is a decomposition of Ω. Suppose that for purposes of diagnosis the doctor carries out a test, with result B. Then formula **5.2.5** gives the posterior probability (i.e. conditional on the result of the test) that the patient is suffering from condition A_j. Here $P(A_j)$ is the prior probability of the same event, i.e. the proportion of patients 'in general' who suffer from A_j. The probability $P(B|A_j)$ is the proportion of the patients suffering from A_j for whom the test will give result B. One would like a completely decisive test, that is, one for which $P(B|A_j)$ was unity for one condition A_j, and zero for all others. However in practice the test will be less conclusive, and one can only hope that it will lead to a substantial sharpening of one's inferences: i.e. that the posterior distribution will be sensibly more concentrated than the prior.

Bayes's theorem has more generally been invoked in cases where one has a number of mutually exclusive hypotheses H_j, and A_j is regarded as the event 'H_j is true', and B as a piece of experimental evidence. Formula **5.2.5** then indicates how the probabilities of truth of the various hypotheses are changed when one takes the experimental evidence into account.

The question is, whether one can meaningfully attach a probability to the event 'H_j is true'. In the medical case one could; one could interpret $P(A_j)$ as the proportion of the population suffering from the jth condition. This was also possible in the lamp example; one interprets $P(A_j)$ as the proportion of lamps on sale which are of manufacture j. But suppose the hypotheses H_j are scientific hypotheses such as, for example, 'the electron is an elementary particle'. In our universe the hypothesis is either true or false (or, possibly, intrinsically meaningless); and there is no conceivable 'sample space' of universes in which such a statement would be sometimes true, sometimes false, and over which one could average. Even if one admitted the idea, there would still be other difficulties: that there is no obvious way to evaluate the $P(A_j)$, or even to ensure that one has an exhaustive collection of hypotheses. Nevertheless, attempts have been made to construct a formal probability theory for such propositions, with $P(A_j)$ interpreted as one's 'degree of belief' in H_j, rather than as a hypothetical proportion. Such an approach has some appeal, but lies off our path.

Exercises

5.2.4 Suppose that $\{A_k\}$ is a decomposition of Ω, and X a r.v. observable on Ω_A, so that the function $X(\omega)$ takes the same value x_k for all ω in A_k. Show that

$$E(X|B) = \frac{\sum_k x_k P(A_k)P(B|A_k)}{\sum_k P(A_k)P(B|A_k)}.$$

5.2.5 Consider the serial number example of exercise 4.3.2. Suppose that the event B is the set of registration numbers observed (repetitions allowed) in a sample of fixed size n, from a town in which cars are registered serially from 1 upwards. Suppose the number of cars in the town is a r.v. $\dot{M}$ (i.e. one is averaging over towns) and one knows in advance that $P(M = m) = \pi_m$. Show that

$$P(M = m \mid B) = \begin{cases} \dfrac{\pi_m m^{-n}}{\sum\limits_{k \geqslant u} \pi_k k^{-n}} & (m \geqslant u), \\ 0 & (m < u), \end{cases}$$

where u is the largest registration number observed in the sample.

5.2.6   Various signals A_j can be sent over a 'noisy' transmission line. The proportion $P(A_j)$ of signals sent which are A_j is known, and the probability $P(B \mid A_j)$ that signal B is received if A_j was sent is also known. One requires a rule for deciding what signal was sent, given that B has been received. Show that the decision rule which minimizes the expected number of mistakes is to say that that signal was sent for which $P(A_j)P(B \mid A_j)$ is maximal, as a function of j.

5.3   A conditional expectation as a random variable

We shall shift ground somewhat in this section, and consider the expectation of one r.v., Y, as being conditioned by the value of another, X.

Suppose then that X takes a countable set of values $\{x_k\}$, and that A_k is the ω-set on which $X(\omega) = x_k$, so that $\{A_k\}$ is a decomposition of Ω, and, by relation **5.1.6**,

$$E(Y) = \sum_k P(A_k)E(Y \mid A_k). \qquad \textbf{5.3.1}$$

But, more generally than this, we can state that

$$E[H(X)Y] = \sum_k P(A_k)H(x_k)E(Y \mid A_k), \qquad \textbf{5.3.2}$$

for the right-hand member of this relation equals

$$\sum H(x_k)E(YI_{A_k}) = E\left[Y \sum_k H(x_k)I_{A_k}\right] = E[YH(X)].$$

(See equation **3.2.3**.)

The statement that relation **5.3.2** holds for any H is the complete converse to the original defining relation **5.1.1**, and equivalent to it. We have just seen that **5.1.1** implies **5.3.2**. By choosing $H(X) = I_{A_j}(\omega)$ in equation **5.3.2**, we see that

$$E(YI_{A_j}) = P(A_j)E(Y \mid A_j), \qquad \textbf{5.3.3}$$

which implies the definition **5.1.1** of $E(Y \mid A_j)$ if $P(A_j) > 0$.

Now, the sum in **5.3.2** is just the expectation of the r.v. $H(X)E(Y \mid X)$, where $E(Y \mid X)$ (a new concept) is a r.v. – that function of X which is equal to $E(Y \mid A_k)$ when $X = x_k$. That is, $E(Y \mid X)$ is obtained from Y by replacing Y by the 'local' conditional average; the average of Y over the set A_k in which ω happens to lie. Thus $E(Y \mid X)$ is a kind of coarsened version of Y, just as Ω_X, the sample space with points A_k, is a coarsened version of Ω.

For example, suppose ω is the census description of an individual, and X is 'year of birth'. Then, if we condense the census description down to a mere specification of X, the original detailed sample space Ω is replaced by one in which 'year of birth' is the only variable, and in replacing any other observable Y on Ω (say, income or number of children) by $E(Y|X)$, we are replacing individual values by average values for a person of the relevant year of birth. Whereas a point ω of Ω represents a fairly individual description, the point A_{30} of Ω_X represents only 'the average thirty year old', for example.

The use of identical notations, $E(Y|A)$ and $E(Y|X)$, for different concepts is unfortunate, but common and convenient. Both quantities are termed 'conditional expectation'; the distinction lies in whether the second argument is an event (when the conditioning is on that *fixed* event) or a r.v. (when the conditioning is on the *random* value of that r.v.).

Theorem 5.3.1

The conditional expectation $E(Y|X)$ is a function of X obeying the relation

$$E[H(X)Y] = E[H(X)E(Y|X)] \qquad \qquad \textbf{5.3.4}$$

for all H. It is equal to that function of X, $G(X)$, which minimizes $E[\{Y-G(X)\}^2]$, at least in the case $E(Y^2) < \infty$.

We have already proved the first assertion: relation **5.3.4** is identical with **5.3.2**, but stated in terms of the r.v. $E(Y|X)$. In this form the two-stage expectation procedure (first an average conditional on X, then an average over X) is quite explicit.

The second idea is a new one, however: that $E(Y|X)$ is that function of X which approximates Y best, in the mean square sense. That is, $E(Y|X)$ is a least square approximant (not linear, in general) of Y in terms of X.

To prove the assertion, denote $E(Y|X)$ by $G(X)$, and any other function of X by $G'(X)$. Then

$$
\begin{aligned}
E[(Y-G')^2] &= E[(Y-G+G-G')^2] \\
&= E[(Y-G)^2] + E[(Y-G)(G-G')] + E[(G-G')^2] \\
&= E[(Y-G)^2] + E[(G-G')^2] \\
&\geqslant E[(Y-G)^2],
\end{aligned}
\qquad \qquad \textbf{5.3.5}
$$

so that condition **5.3.4**, which brought about the vanishing of

$$E[(Y-G)(G-G')],$$

is sufficient to ensure that the mean square deviation is minimal when $E(Y|X)$

is the approximant. The condition is also necessary: if for some H, normalized in scale so that $E(H^2) = 1$,

$$E(HY) - E(HG) = \Delta \neq 0, \qquad \qquad \text{5.3.6}$$

then

$$E[(Y - G - \Delta H)^2] = E[(Y - G)^2] - \Delta^2 < E[(Y - G)^2], \qquad \text{5.3.7}$$

so that $G + \Delta H$ would be a better approximant to Y than G is.

The final inequality of **5.3.5** will be strict unless $G' \overset{\text{m.s.}}{=} G$, when we readily find that $E(Y|X) = G'(X)$ also satisfies **5.3.4**. This indeterminacy in the determination of $E(Y|X)$ arises when $P(A_j)$ is zero for some j, when we see from **5.3.3** that $E(Y|A_j)$ is indeterminate in value. Indeed, the value assigned in such a case is immaterial, because all that is operationally required of a conditional expectation is that it should satisfy relation **5.3.4**.

Setting $G' = 0$ in **5.3.5**, we see that

$$E(Y^2) = E[(Y - G)^2] + E(G^2). \qquad \qquad \text{5.3.8}$$

Thus, if $E(Y^2) < \infty$, we know that $E[(Y - G)^2] < \infty$, and inequality **5.3.5** is meaningful. If $E(Y^2)$ is infinite, then the least square characterization is not so natural, although the idea can be adapted.

Exercise

5.3.1   Note that in the present notation we could write the conclusion of exercise 5.1.1 as

$$E(R_2|R_1) = \rho V_2 + \frac{V_0}{V_1}(R_1 - \rho V_1),$$

and that this indeed is a r.v., and a function of R_1.

5.4   Conditioning on a σ-field of events

We encounter conditioning events of zero probability in an inescapable fashion as soon as we consider the situation of the previous section, where the distribution of one r.v., Y, is conditioned by the value of another, X, but in the case where X is continuously distributed. For example, we might wish to investigate the distribution of wind-speed for winds in a given direction, or of the maximum height reached by a sounding rocket (which means, the distribution of rocket height, Y, conditioned by the fact that the rate of change of height, X, is equal to zero).

One's natural approach is to consider the conditioning event $X = x$ as the limit of the event $x - h < X < x + h$ as h tends to zero. One's next discovery is then that, in this sense, the event $X = x$ is not equivalent to the event $X/Y = x/Y$, for example. For, suppose that X and Y have joint density $f(x, y)$, so that

$$E[H(X, Y)] = \int \int H(x, y) f(x, y) \, dx \, dy$$

for a suitably large class of functions H; then, by definition **5.1.1**,

$$E\left[H(Y)|(|X - x| < h)\right] = \frac{\displaystyle\int_{x-h}^{x+h} du \int dy \, H(y) f(u, y)}{\displaystyle\int_{x-h}^{x+h} du \int dy \, f(u, y)},$$

which means that, in the limit of small h, the r.v. has conditional density

$$f(y | X = x) = \frac{f(x, y)}{\displaystyle\int f(x, y) \, dy}, \qquad\qquad \textbf{5.4.1}$$

at least if $f(x, y)$ is continuous.

Now, the event $x Y^{-1} - h < X Y^{-1} < x Y^{-1} + h$ is equivalent to

$$x - h|Y| < X < x + h|Y|,$$

so that, by a derivation analogous to that of **5.4.1**, we obtain

$$f\left(y \left| \frac{X}{Y} = \frac{x}{Y} \right.\right) = \frac{f(x, y)|y|}{\displaystyle\int f(x, y)|y| \, dy}, \qquad\qquad \textbf{5.4.2}$$

which indeed differs from **5.4.1**.

In fact an event such as $X = x$ must be embedded in a field of events if it is to be clearly specified: in this case the field is that generated from events of the type $X \leqslant x$.

The way adopted to clarify the situation is to abandon the elementary constructive definition **5.1.1**, and to appeal to relation **5.3.4** as the essential defining property of $E(Y|X)$. That is, $E(Y|X)$ is *defined* as a function of X that satisfies **5.3.4** for all H of a class of functions that includes at least the indicator functions of intervals. In the case where X assumes at most countably many values, this requirement implies the elementary definition, as we saw from **5.3.3**.

We leave it as an exercise for the reader to show that, if X and Y have a joint density, then formula **5.4.1** for the conditional density of Y given X is indeed consistent with this definition.

The first assertion of Theorem 5.3.1 has now the status of a definition rather than a deduction. The second assertion still holds in this more general case, however. This extremal characterization of $E(Y|X)$ has the valuable property that it ensures that equation **5.4.4** is self-consistent, and really possesses a solution. This is because there must *be* a function $G(X)$ that minimizes

$E[\{Y-G(X)\}^2]$, at least in the sense of existing as the limit of a sequence of functions $\{G_n(X)\}$ for which $E[\{Y-G_n(X)\}^2]$ approaches its lower bound as n increases (the sequence $\{G_n\}$ necessarily being convergent in the m.s. sense: see exercise 5.4.3), and we saw in the previous section that $G(X) = E(Y|X)$ then necessarily obeyed relation **5.3.4**.

Note that the idea of approximating $E(Y|X)$ by a sequence of 'elementary' conditional expectations is something we have already tried, when we took the conditioning event as being $x-h < X < x+h$ rather than $X = x$. In effect, instead of taking the conditioning field $\mathscr{B}$ as being that generated from the events $\{X \leqslant x\}$, we took it as the field $\mathscr{B}_n$ made up of the events

$$\frac{j}{n} < X \leqslant \frac{j+1}{n} \quad (j \text{ integral}).$$

We could obviously just as well condition by the values of several r.v.s as of one; the formal generalization will be plain. More generally, if the conditioning events form a general field $\mathscr{B}$, then what we have denoted $E(Y|X)$ is written $E^{\mathscr{B}}(Y)$, and is defined to be a $\mathscr{B}$-measurable r.v. satisfying

$$E(YZ) = E[ZE^{\mathscr{B}}(Y)] \qquad\qquad\qquad \textbf{5.4.3}$$

for any $\mathscr{B}$-measurable r.v. Z. However, there is little point in studying more general cases before one has become familiar with the simpler ones.

Exercises

5.4.1 Consider a distribution uniform on the surface of the unit sphere, in that the probabilitȳ that a sample point lies in a given region is proportional to the area of that region. The sample point will have Cartesian coordinates,

$$\left. \begin{array}{l} x = \cos\phi\cos\theta \\ y = \cos\phi\sin\theta \\ z = \sin\phi \end{array} \right\} \quad (-\tfrac{1}{2}\pi \leqslant \phi \leqslant \tfrac{1}{2}\pi; 0 \leqslant \theta < 2\pi),$$

if ϕ and θ are its angles of latitude and longitude. Show that, if the point is constrained to lie on the meridian $\theta = 0$, then its ϕ coordinate has conditional density $\tfrac{1}{2}\cos\phi$ or $1/\pi$ according as the meridian is specified by 'θ arbitrarily small' or 'y arbitrarily small'.

5.4.2 Suppose the minimal value of $E[\{Y-G(X)\}^2]$ with respect to the function G is Δ; we shall also denote the minimizing function itself by $G(X)$. Suppose that for an approximation $G_n(X)$ to the minimizing function we have

$$E[\{Y-G_n(X)\}^2] = \Delta + \varepsilon_n^2.$$

Show that $E[(G_n-G)^2] = \varepsilon_n^2$, and that equation **5.3.4** is approximately fulfilled by G_n in the sense that

$$|E(HY) - E(HG_n)| \leqslant \varepsilon_n \sqrt{[E(H^2)]}.$$

5.4.3 Show that, for the case of exercise 5.4.2, $E[(G_m - G_n)^2] \leqslant \varepsilon_m^2 + \varepsilon_n^2$. That is, if $\varepsilon_n \to 0$ then $E[(G_m - G_n)^2] \to 0$. We shall see from Theorem 9.7.2 that this implies that the sequence $\{G_n\}$ has a limit G in the mean square sense.

5.5 Statistical independence

Some r.v.s are obviously related: for example, the height and weight of a randomly chosen subject. However, one feels that in some other cases there should be no relation at all. For instance, one would expect that the outcome of a second throw of a die is not related to the outcome of the first, and that the deviation of today's temperature from its seasonal mean bears no relation to the corresponding deviation several years ago.

We shall now give a formal definition and say that two observables X and Y (not necessarily numerically valued) are *statistically independent* if

$$E[H(X)K(Y)] = E[H(X)]E[K(Y)] \qquad\qquad \textbf{5.5.1}$$

for all functions H and K such that the expectations in the right-hand member exist.

Although reasonable as a criterion, condition **5.5.1** may not obviously express one's intuitive idea of independence, which is, roughly, that 'knowledge of X does not help one to infer the value of Y'. To see that it does, choose $H(X)$ as the indicator function $I_A(X)$ of an ω-set $X(\omega) \in A$, so that condition **5.5.1** becomes

$$E[I_A(X)K(Y)] = E[I_A(X)]E[K(Y)], \qquad\qquad \textbf{5.5.2}$$

$$\text{or} \quad E[K(Y) | X \in A] = E[K(Y)]. \qquad\qquad \textbf{5.5.3}$$

That is, $K(Y)$ has the same average value whether one takes the over-all expectation, or an expectation conditioned by the fact that $X(\omega)$ is confined to one of the sets A of Ω_X. The knowledge that X lies in A thus has no predictive value for Y, since the expectation value of $K(Y)$ is unaffected by it.

For example, X and Y might be one's scores on the first and second throws of a die. Independence of X and Y would imply that $E[K(Y) | X = j]$ has the same value for $j = 1, 2, \ldots, 6$, which is just the statement that X has no predictive value for Y.

One could start with condition **5.5.3** as one's criterion of independence, for appropriate A, and deduce **5.5.1** from it (see exercise 5.5.7). However, condition **5.5.1** is obviously simpler; it is symmetric in X and Y, it avoids the sometimes delicate concept of conditioning altogether, and yields the second characterization **5.5.3** immediately. All these advantages are probably related to the fact that **5.5.1** is the operational expression of independence, the actual property to which one appeals in applications.

By taking $K(Y) = I_B(Y)$ in **5.5.2**, we see that

$$P(X \in A, Y \in B) = P(X \in A)P(Y \in B) \qquad\qquad \textbf{5.5.4}$$

for A and B sets of Ω_X and Ω_Y respectively.

The generalization of condition **5.5.1** to the case of several observables is immediate; we say that observables $X_1, X_2, \ldots, X_n$ are statistically independent if

$$E\left[\prod_1^n H_j(X_j)\right] = \prod_1^n E[H_j(X_j)] \qquad\qquad \textbf{5.5.5}$$

for all functions H_j for which the right-hand member of **5.5.5** is defined.

Exercises

5.5.1 Show that the observables $\xi_1, \xi_2, \ldots, \xi_n$ of equations **4.3.1** and **4.7.1** are independent, but that these observables are not independent in the context of section 4.5.

5.5.2 If the observable X corresponds simply to the occurrence or non-occurrence of an event A, then $H(X)$ can only be of the form $\alpha + \beta I_A(\omega)$, where α and β are constants. Show that if Y also corresponds to the occurrence or non-occurrence of an event B, then condition **5.5.1** is equivalent to the single relation

$$P(AB) = P(A)P(B). \qquad\qquad \textbf{5.5.6}$$

In this case the *events* A and B are said to be independent. Relation **5.5.6** is sometimes referred to as the 'multiplicative law of probabilities' and coupled with the 'additive law' (relation **3.1.3**) in this respect. However, the status of the two are different: whereas the additive law is either an *axiom* or an immediate consequence of one, the multiplicative law is a consequence of the *definition* of independence.

5.5.3 Suppose that the observable X_j can only take a denumerable set of values, and denote the event that it takes the kth value by

$$A_{jk} \quad (j = 1, 2, \ldots, n; k = 1, 2, \ldots).$$

Show that in this case condition **5.5.5** is equivalent to

$$P\left(\bigcap_{j=1}^n A_{jk_j}\right) = \prod_{j=1}^n P(A_{jk_j}) \qquad\qquad \textbf{5.5.7}$$

for any $k_1, k_2, \ldots, k_n$. Thus, the r.v.s $N_1, N_2, \ldots, N_r$ of **4.6.5** are independent.

5.5.4 Suppose that the observables X_j can only take positive integral values. Show that condition **5.5.5** is then equivalent to

$$E\left(\prod_1^n z_j^{X_j}\right) = \prod_1^n E(z_j^{X_j}) \qquad\qquad \textbf{5.5.8}$$

for $|z_1| = |z_2| = \ldots = |z_n| = 1$, and hence that **4.7.12** implies independence of the r.v.s R_k. This and the previous example show that in certain cases one need demand condition **5.5.5** only for a particular restricted set of functions H_j. (Respectively, $I_{A_{jk}}$ for $k = 1, 2, \ldots,$ and $z_j^{X_j}$ for $|z_j| = 1$.)

5.5.5   Show that condition **5.5.5** implies that

$$E\big[H(X_j)\,|\, X_k \in A_k, k \neq j\big]$$

does not depend upon the A_k; that is, $n-1$ of the observables have collectively no predictive value for the nth.

5.5.6   Observables can be pairwise independent without being independent. To see this, consider the example below, of three r.v.s defined on a sample space of four points.

ω	ω_1	ω_2	ω_3	ω_4
$P(\omega)$	$\frac{1}{4}$	$\frac{1}{4}$	$\frac{1}{4}$	$\frac{1}{4}$
X_1	1	0	0	1
X_2	0	1	0	1
X_3	0	0	1	1

5.5.7   Show that if $E\big[K(Y)\,|\,X\big]$ is constant (as a function of X) then it must necessarily equal $E[K(Y)]$, and that substitution of this assertion in the defining relation **5.3.4** for a conditional expectation yields condition **5.5.1**.

5.5.8   Suppose that r.v.s X and Y are independent, and that $E[K(Y)^2] < \infty$. Show that if $E\big[K(Y)\,|\,X\big]$ is any solution of the defining relation for the conditional probability, then $E\big[K(Y)\,|\,X\big] \overset{\text{m.s.}}{=} E[K(Y)]$. In fact, one would simply take the constant $E[K(Y)]$ as the obvious sensible solution.

5.6   Elementary consequences of independence

We have already remarked (exercise 5.5.1) that the 'equiprobability' assumption for the sampling sequences of sections 4.2, 4.3 and 4.7 amounts to an assumption of independence. We can now make this assumption explicitly.

For example, suppose that one has a succession of n trials, in each of which one can register 'success' or 'failure'. Suppose the outcomes of the trials are statistically independent, and the probability of a success is p in each case. Then, by equation **5.5.7**, the probability that the sequence contains r successes, in some given order, is just $p^r q^{n-r}$ (with $q = 1-p$, as always), and the probability of r successes, order immaterial, is then given by the binomial distribution:

$$P_r = \binom{n}{r} p^r (1-p)^{n-r}. \qquad\qquad \textbf{5.6.1}$$

This was just the result we found from equation **4.3.1** for the probability that

r out of n draws from a pack would be red, with $p = m_1/m$, the proportion of red cards in the pack. Now we have the same distribution in a more general context, based on the assumption of independence of trials, rather than on that of some underlying set of equiprobable outcomes. The success probability p may take any value in $[0, 1]$.

Such a sequence of independent trials is known as a sequence of *Bernoulli trials*. For such trials all the distributions derived in chapter 4 on the basis of equation **4.3.1** are valid: the binomial distribution **4.3.2** for the total number of successes in n trials; the geometric distribution **4.6.1** for the number of trials required to reach the first success; the negative binomial distribution **4.6.8** for the number of trials required to reach the rth success. For the continuum case of section 4.7, formula **4.7.1** implies that the observables are independent r.v.s. In the Poisson limit we are essentially considering points which are uniformly and independently distributed over an infinite region T at a given average density ρ.

The behaviour of sums of independent r.v.s provides a classical and rewarding study. Suppose (for simplicity rather than necessity) that r.v.s X and Y can take only non-negative integral values. Then, by the additive law for probabilities,

$$P(X + Y = r) = \sum_j P(X = j, Y = r-j). \qquad 5.6.2$$

Suppose, however, that X and Y are independent. Then **5.6.2** must simplify to

$$P(X + Y = r) = \sum_j P(X = j)P(Y = r-j). \qquad 5.6.3$$

Relation **5.6.3** takes a more revealing form if we consider the p.g.f.s of the r.v.s concerned. We use the general notation $\Pi_\xi(z) = E(z^\xi)$; for the p.g.f. of a r.v. ξ.

Theorem 5.6.1

If X and Y are independent integral-valued r.v.s then

$$\Pi_{X+Y}(z) = \Pi_X(z)\Pi_Y(z). \qquad 5.6.4$$

The result follows simply from the fact that

$$E(z^{X+Y}) = E(z^X z^Y) = E(z^X)E(z^Y) \qquad 5.6.5$$

(cf. condition **5.5.8**). However, it is one of the most important basic identities of probability theory. For example, relation **5.6.3** follows immediately if we equate coefficients of z^r in the two members of **5.6.4**.

As an example, suppose that X and Y are independent Poisson variables with parameters λ and μ respectively. Then by **5.6.4** $X + Y$ will have p.g.f. $\exp[(\lambda+\mu)(z-1)]$, so that $X + Y$ is also a Poisson variable, with parameter $\lambda+\mu$.

Suppose we say that a r.v. with the binomial distribution **5.6.1** is a $b(n, p)$ variable. Then, we deduce in the same way that, if X and Y are independent r.v.s, $b(m, p)$ and $b(n, p)$ respectively, then $X + Y$ is a $b(m+n, p)$ variable (as is clear, in fact, from the derivation of the binomial distribution).

Relation **5.6.4** extends immediately to the case of several summands. If

$$S = \sum_{1}^{n} X_j, \qquad\qquad\qquad \textbf{5.6.6}$$

and the X_j are independently distributed, then

$$\Pi_S(z) = \prod_{1}^{n} \Pi_{X_j}(z). \qquad\qquad\qquad \textbf{5.6.7}$$

In particular, if the X_j are identically distributed, then

$$\Pi_S(z) = [\Pi_X(z)]^n, \qquad\qquad\qquad \textbf{5.6.8}$$

where Π_X is the p.g.f. of an individual X.

The p.g.f. of a $b(n, p)$ variable

$$\Pi(z) = (pz + q)^n \qquad\qquad\qquad \textbf{5.6.9}$$

provides a special case of this: the variable 'number of successes' is of the form **5.6.6**, where the X_j are independent and take the values 1 or 0 with probabilities p and q.

Exercises

5.6.1   Consider a sequence of trials in which the outcomes of each trial are independent, and any trial can have outcome k with probability p_k ($k = 1, 2, \ldots, s$). Let N_k be the number of times outcome k is observed in n trials. Show that the joint distribution of $N_1, N_2, \ldots, N_s$ is just the multinomial distribution **4.4.4**.

5.6.2   Interpret the p.g.f. **4.6.7** of a negative binomial variable in the light of equation **5.6.8**.

5.6.3   *Controlled overbooking.* Suppose an airline allows n passengers to book on a flight with maximum capacity m ($n \geqslant m$). Each passenger who flies brings the line revenue a; each passenger for whom there is no place at flight costs the line an amount b. Passengers have a probability $1 - p$ of cancelling their flight, independently of one another, so that X, the number of passengers finally wishing to fly, is $b(n, p)$. Show that the value of n maximizing the expected revenue is the smallest value for which

$$P(X < m \mid n) \leqslant \frac{b}{a+b}.$$

This looks similar to the solutions of section 2.5, but the independence assumption is essential to this particular problem.

Consider two independent Poisson variables X and Y with parameters λ and μ; let us calculate the distribution of X conditional on the value of $X + Y$. We have, by **5.1.10**,

$$P(X = r, Y = n-r \mid X + Y = n) = \frac{P(X = r)P(Y = n-r)}{P(X + Y = n)}$$

$$= \frac{e^{-\lambda}(\lambda^r/r!)e^{-\mu}\mu^{n-r}/(n-r)!}{e^{-(\lambda+\mu)}(\lambda+\mu)^n/n!}$$

$$= \binom{n}{r}p^r(1-p)^{n-r}, \qquad\qquad \textbf{5.6.10}$$

where $p = \lambda/(\lambda+\mu)$. That is, the conditional distribution of X is binomial, which is reasonable if we consider the basis on which the Poisson distribution was derived in section 4.7. Note that we have already proved this result by direct use of p.g.f.s, after equation **5.1.5**.

The difference between formula **5.6.10** and the unconditional result

$$P(X = r, Y = s) = \frac{e^{-(\lambda+\mu)}\lambda^r\mu^s}{r!s!} \qquad\qquad \textbf{5.6.11}$$

is related to the idea of a *stopping rule*. Suppose that one is interested in the relative frequency of two types of insect in the country, and that, on making a survey, one counts X of one type and Y of the other. If the stopping rule for the survey is not in terms of X or Y (e.g. 'sample a given area', or 'continue sampling for three hours') then it is plausible that X and Y are independent Poisson variables, and that distribution **5.6.11** is valid (if insects occur independently and uniformly in space or time, as did the molecules of section 4.7). However, suppose the rule is 'continue until a total of n insects has been gathered'. Then distribution **5.6.11** must be modified by the constraint $X + Y = n$, and it is this conditioning that yields formula **5.6.10**. Yet another stopping rule would be 'continue until exactly r insects of the first type have been gathered'. This leads to the negative binomial distribution of section 4.6.

Consider the same situation for a pair of binomial variables. If X and Y are independent, and are $b(m_1, p)$ and $b(m_2, p)$ respectively, then we find that

$$P(X = r \mid X + Y = n) = \frac{\binom{m_1}{r}p^r q^{m_1-r}\binom{m_2}{n-r}p^{n-r}q^{m_2-n+r}}{\binom{m}{r}p^n q^{m-n}}$$

$$= \binom{m_1}{r}\binom{m_2}{n-r}\Bigg/\binom{m}{n}, \qquad\qquad \textbf{5.6.12}$$

where $m = m_1 + m_2$. That is, X follows the hypergeometric distribution

4.5.1. If we consider this example in terms of the population-sampling problem discussed in section 4.5 we see again that it is a question of specifying a stopping rule. Suppose that the second sample was taken from the lake in such a way that any given fish, marked or unmarked, had probability p of capture, independent of what happened to other fish. The numbers X and Y of marked and unmarked fish caught in the second sample would then be independent binomial variables, as indicated. However, suppose that sampling is prolonged until exactly n fish have been caught. The constraint $X + Y = n$ then leads one to the hypergeometric distribution **5.6.12**.

It is convenient that expression **5.6.12** is independent of p. Note that we would not have had this cancellation had the success probability not been the same for X and Y, i.e. if marked and unmarked fish had had different individual probabilities of capture.

Consider yet another variety of problem, in which we perform the operation inverse to conditioning: averaging over an additional r.v. Suppose that the number of insects in an area is a Poisson r.v. X with expectation λ. An entomologist is searching the area, and has a chance p of observing any given insect, this event being independent for different insects. We are interested in the distribution of Y, the number of insects actually observed. Conditional on X, the r.v. Y must be $b(X, p)$. Averaging over X, as we did over k in formula **5.1.6**, we find that the unconditioned distribution of Y is

$$
P(Y = r) = \sum_{n=0}^{\infty} P(X = n)P(Y = r \mid X = n)
$$

$$
= \sum_{n=r}^{\infty} \frac{e^{-\lambda}\lambda^n}{n!} \binom{n}{r} p^r(1-p)^{n-r}
$$

$$
= \frac{e^{-\lambda p}(\lambda p)^r}{r!}. \qquad\qquad \textbf{5.6.13}
$$

Thus Y is a Poisson r.v. with expectation λp, which agrees with intuition.

Exercises

5.6.4 Generalize calculation **5.6.10** to show that independent Poisson variables are multinomially distributed if their sum is constrained.

5.6.5 Generalize **5.6.12** similarly.

5.6.6 Suppose that $N_1 + 1, N_2 + 1, \ldots, N_s + 1$ are independent geometric variables, with common expectation. Show that the distribution of $N_1, N_2, \ldots, N_s$ conditional on $\sum N_k = n$ is the Bose–Einstein distribution of section 4.9.

5.6.7 Show that the r.v. R with distribution **5.6.12** has p.g.f. $\Pi(z)$ proportional to the coefficient of w^n in $(pwz + q)^{m_1}(pw + q)^{m_2}$, and hence derive the moment formula **4.5.4**.

5.6.8   We know that, for a sequence of Bernoulli trials, the number of trials until the rth success (r prescribed) has a negative binomial distribution. Why do we not then find that, corresponding to **5.6.10**, the distribution of $X + Y$ conditional on $X = r$ is negative binomial, if X and Y are independent Poisson variables?

5.6.9   Suppose that, in the insect example last mentioned in the text, X had a general distribution with p.g.f. $\Pi_X(z)$. Show that

$$\Pi_Y(z) = \Pi_X(pz + q).$$

As well as formulating a natural and essential idea, the independence concept is a powerful tool for building up interesting processes from simple elements. For instance, in the case of a Bernoulli sequence of trials, we need only specify expectations on the individual sample space Ω_j of outcomes of the jth trial. The independence concept allows us to consider expectations on the composite sample space

$$\Omega = \Omega_1 \times \Omega_2 \times \dots$$

in which the elementary outcomes are the composite sequences of individual outcomes (cf. exercise 4.6.6, formula **6.3.1**).

With the formalization of the independence concept we can now attack problems of genuine scientific interest. The theory of Mendelian inheritance provides examples of models which can be simple, without being idealized to the point where they have no practical value.

The gene is the unit of heredity, and in the simplest case genes occur in pairs: each gene of a particular pair can assume two forms (alleles), A and a. There are, then, with respect to this gene, three types of individual (*genotypes*): AA, Aa and aa. The pure genotypes AA and aa are termed *homozygotes*, the mixed one Aa a *heterozygote*. If A *dominates* a then the heterozygote Aa will be outwardly indistinguishable from the homozygote AA; there are then only two outwardly distinguishable types (*phenotypes*): (AA or Aa) and aa. For example, brown eyes are dominant over blue, so an individual with a 'blue' and a 'brown' gene will have brown eyes. (The situation is actually more complicated, but such a model is a first approximation to it.)

An individual receives one gene at random from each of its parents; i.e. it receives a given maternal gene and a given paternal gene with probability $\frac{1}{4}$, for each of the four possibilities. Thus, the mating $AA \times Aa$ would yield progeny of types AA or Aa, each with probability $\frac{1}{2}$: the mating $Aa \times Aa$ would yield AA, Aa or aa, with respective probabilities $\frac{1}{4}, \frac{1}{2}$ and $\frac{1}{4}$.

Suppose now that we have a large population, and that the proportions of the genotypes AA, Aa and aa in the nth generation are p_n, q_n and r_n respectively. The point of assuming the population large is that p_n can then be equated with the probability that a randomly chosen individual is an AA, p_n^2 with the probability that a randomly chosen *pair* are both AA, etc.

Assume now that mating takes place at random, i.e. the two parents are chosen randomly and independently from the population. Thus, the probability of an $AA \times Aa$ mating in the nth generation would be $2p_n q_n$, etc. The probability that the progeny is an AA is then

$$p_{n+1} = p_n^2 + 2p_n q_n(\tfrac{1}{2}) + q_n^2(\tfrac{1}{4})$$

$$= (p_n + \tfrac{1}{2}q_n)^2 = \theta_n^2, \qquad\qquad \textbf{5.6.14}$$

say, where θ_n is the proportion of A genes in the nth generation. We leave it to the reader to verify similarly that

$$q_{n+1} = 2\theta_n(1 - \theta_n),$$
$$\qquad\qquad\qquad\qquad\qquad\qquad\qquad \textbf{5.6.15}$$
$$r_{n+1} = (1 - \theta_n)^2,$$

and that

$$\theta_{n+1} = p_{n+1} + \tfrac{1}{2}q_{n+1} = \theta_n. \qquad\qquad \textbf{5.6.16}$$

That is, the gene frequency θ_n stays constant from generation to generation (at θ, say), and, after one generation of random mating, the genotype frequencies become fixed at θ^2, $2\theta(1 - \theta)$ and $(1 - \theta)^2$. Hence,

$$4pr = q^2$$

(the *Hardy–Weinberg law*).

Exercise

5.6.10   If A is dominant, and aa is regarded as 'abnormal', show that the probability that the first child of normal parents is abnormal is $\{(1 - \theta)/(2 - \theta)\}^2$. (We assume that mating has been random in previous generations.)

5.7   **Partial independence; orthogonality**

We must live with the idea that we may know $E(X)$ only for certain X, or that for a given r.v. X we may know $E[H(X)]$ only for certain H. Correspondingly, for a given pair of r.v.s X and Y, we may be able to assert the validity of condition **5.5.1** only for certain H and K. In such a case X and Y have only a partial degree of independence.

An extreme case of this is that in which we can assert that validity of **5.5.1** only for H and K linear, which means, essentially, that we know only that

$$E(XY) = E(X)E(Y). \qquad\qquad \textbf{5.7.1}$$

This relation can, in virtue of **2.6.7**, be written

$$\text{cov}(X, Y) = 0. \qquad\qquad \textbf{5.7.2}$$

Such a pair of r.v.s is said to be *uncorrelated*. In the special case when the means are zero, so that relation **5.7.1** becomes $E(XY) = 0$, they are said to be *mutually orthogonal*. The concept of lack of correlation, or orthogonality is important, because it is the nearest one can come to the concept of independence if one is restricted to a knowledge of second moments, as in section 2.5.

Just as independence means that X has no predictive value for Y, so lack of correlation means that X has no predictive value for Y *in the linear least square sense*. That is, suppose we consider a predictor for Y which is linear in X:

$$\hat{Y} = \alpha + \beta X, \qquad\qquad 5.7.3$$

and choose α and β so as to minimize $E[(\hat{Y} - Y)^2]$. Then we find that the optimal values of β and prediction mean square error are

$$\beta = \frac{\text{cov}(X, Y)}{\text{var}(X)}, \qquad\qquad 5.7.4$$

$$E[(\hat{Y} - Y)^2] = \text{var}(Y) - \frac{[\text{cov}(X, Y)]^2}{\text{var}(X)}. \qquad\qquad 5.7.5$$

Thus, in the case **5.7.2**, the r.v. X receives zero coefficient in the prediction formula **5.7.3**, and $E[(\hat{Y} - Y)^2]$ has its maximal value of $\text{var}(Y)$ – just the value it would have if we had used the constant $E(Y)$ as predictor of Y.

That the property **5.7.2** is genuinely weaker than independence is shown by the example $X = \sin\theta$, $Y = \cos\theta$, with θ uniformly distributed on $(-\frac{1}{2}\pi, \frac{1}{2}\pi)$. Here X and Y are uncorrelated, and so X has no linear predictive value for Y. However, the two variables are not independent, since an exact nonlinear predictor exists: $Y = \sqrt{(1 - X^2)}$.

A case intermediate between lack of correlation and independence is that in which equation **5.5.1** holds only for linear K, so that

$$E[H(X)Y] = E(Y)E[H(X)] \qquad\qquad 5.7.6$$

for any H for which $E[H(X)]$ is defined. Relation **5.7.6** is equivalent to

$$E(Y|X) = E(Y), \qquad\qquad 5.7.7$$

so that an X-conditioning does not affect the expectation of Y, although it might affect the expectation of other functions of Y.

Exercises

5.7.1 Consider the *correlation coefficient*

$$\rho = \frac{\text{cov}(X, Y)}{[\text{var}(X)\text{var}(Y)]^{\frac{1}{2}}}.$$

We know from the Cauchy inequality **2.7.10** that $\rho^2 \leqslant 1$. Show that

$$\frac{\min_{\alpha,\beta} E(Y - \alpha - \beta X)^2}{\min_{\alpha} E(Y - \alpha)^2} = 1 - \rho^2,$$

so that ρ^2 provides a dimensionless measure of the degree of linear dependence between X and Y. For measures of *general* dependence, see exercise 12.2.3.

5.7.2   Show that if $X_1, X_2, \ldots, X_n$ are mutually uncorrelated, then

$$\text{var}\left(\sum a_j X_j\right) = \sum a_j^2 \, \text{var}(X_j),$$

and $\text{cov}\left(\sum a_j X_j, \sum b_j X_j\right) = \sum a_j b_j \, \text{var}(X_j).$

Here the a_j and b_j are constant coefficients.

5.7.3   Suppose that Y_n is the linear least square predictor of Y in terms of $X_1, X_2, \ldots, X_n$. Show that $Y - Y_n$ is orthogonal to $X_1, X_2, \ldots, X_n$ (cf. section 2.6) and hence that $\Delta_1 = Y_1$, $\Delta_2 = Y_2 - Y_1$, $\Delta_3 = Y_3 - Y_2, \ldots,$ $\Delta_n = Y_n - Y_{n-1}$, $Y - Y_n$ are mutually orthogonal, so that $\sum E(\Delta_j^2) \leqslant E(Y^2)$.

5.7.4   Suppose that $Y_n = E(Y \mid X_1, X_2, \ldots, X_n)$ is the general least square predictor of Y in terms of $X_1, X_2, \ldots, X_n$ (see Theorem 5.3.1). Show that $E(Y - Y_n \mid X_1, X_2, \ldots, X_n) = E(Y - Y_n) = 0$, but that $Y - Y_n$ is not necessarily independent of $X_1, X_2, \ldots, X_n$. Adopting the notation of exercise 5.7.3, show that $\Delta_0 = E(Y)$, $E(\Delta_j \mid \Delta_1, \ldots, \Delta_{j-1}) = E(\Delta_j) = 0$ $(j > 0)$, and that $\sum E(\Delta_j^2) \leqslant E(Y^2)$.

6 Applications of the Independence Concept

6.1 Mean square convergence of sample averages

Suppose r.v.s $X_1, X_2, \ldots, X_n$ possess variances. Then the variance of a linear form $\sum a_j X_j$ is given by

$$\text{var}(\textstyle\sum a_j X_j) = E[\textstyle\sum a_j \{X_j - E(X_j)\}]^2$$
$$= \textstyle\sum\sum a_j a_k E[\{X_j - E(X_j)\}\{X_k - E(X_k)\}]$$
$$= \textstyle\sum\sum a_j a_k \text{cov}(X_j, X_k). \qquad\qquad \textbf{6.1.1}$$

In the particular case of uncorrelated variables (and, *a fortiori*, in the case of independent variables) this formula reduces to

$$\text{var}(\textstyle\sum a_j X_j) = \textstyle\sum a_j^2 \, \text{var}(X_j). \qquad\qquad \textbf{6.1.2}$$

Suppose now that $X_1, X_2, \ldots$ (still supposed uncorrelated) have a common expectation $E(X_j) = \mu$, so that they can be regarded as observations on an unknown quantity μ. For example, they might be repeated measurements of the mass of the electron. We shall not necessarily suppose that the measurements are equally precise, and so shall set $\text{var}(X_j) = v_j$.

Consider now the sample mean

$$\overline{X}_n = \frac{1}{n} \sum_1^n X_j \qquad\qquad \textbf{6.1.3}$$

for which $E(\overline{X}_n) = \mu$, and, by formula **6.1.2**,

$$E[(\overline{X}_n - \mu)^2] = \text{var}(\overline{X}_n) = \frac{1}{n^2} \sum_1^n v_j. \qquad\qquad \textbf{6.1.4}$$

Thus, if the v_j are such that

$$\frac{1}{n^2} \sum_1^n v_j \to 0 \quad \text{as} \quad n \to \infty \qquad\qquad \textbf{6.1.5}$$

(as would be the case if the v_j were constant or bounded, for example), then we see from **6.1.3** that

$$\overline{X}_n \xrightarrow{\text{m.s.}} \mu. \qquad\qquad \textbf{6.1.6}$$

That is, the sample mean converges in mean square to the expectation value μ. This conclusion, generalizing that of **4.2.7**, is very important, because our whole theory was based on the empirical finding that a sample average such as $\overline{X}_n$ seems to tend to a 'limit' as n increases, and on the postulate that the expectation $E(X)$ is the idealization of this 'limit'. Now we find that this convergence can be reproduced within our theory, and can be given the precise sense of **6.1.6**.

As before, an application of Chebichev's inequality shows that

$$P(|\overline{X}_n - \mu| > \varepsilon) \leqslant \frac{\text{var}(X_n)}{\varepsilon^2} \to 0 \qquad \qquad \textbf{6.1.7}$$

for arbitrarily small fixed ε. Relation **6.1.7** states that $\overline{X}_n$ *converges to μ in probability*; this is always a consequence of mean square convergence.

We obtained the result **6.1.6** on the assumptions that the X_j were uncorrelated, and that **6.1.5** held. These assumptions can be weakened (cf. section 6.2), but it is plain that some such conditions are required – namely, that the observations should show some degree of independence, and that they should not be too imprecise. In brief, new observations should bring new information, and they should bring *enough* new information.

The convergence concept is one that will recur repeatedly: the most general conclusions of probability theory take the form of limit theorems.

If one regards one's task as the 'estimation' of μ from $X_1, X_2, \ldots, X_n$, then one can in general improve on the estimate $\overline{X}_n$ by giving relatively greater weight to the more precise observations. Thus, suppose we consider a linear estimate

$$\hat{\mu}_n = \sum_1^n a_j X_j, \qquad \qquad \textbf{6.1.8}$$

and choose the coefficients a_j to minimize $E[(\hat{\mu}_n - \mu)^2]$, subject to the restriction that $E(\hat{\mu}_n) = \mu$ identically in μ (the restriction of *unbiasedness*). We leave it to the reader to show that the optimal estimate thus obtained is

$$\mu_n^* = \frac{\sum_1^n (X_j/v_j)}{\sum_1^n (1/v_j)}, \qquad \qquad \textbf{6.1.9}$$

and that

$$E[(\mu_n^* - \mu)^2] = \left[\sum_1^n (1/v_j) \right]^{-1} \qquad \qquad \textbf{6.1.10}$$

Thus $\mu_n^* \xrightarrow{\text{m.s.}} \mu$ iff $\sum_1^\infty v_j^{-1} = \infty$, which is a weaker condition than **6.1.5**.

Exercises

6.1.1   It is not necessary that the observations be completely uncorrelated for equation **6.1.6** to be valid: one requires merely that $n^{-2} \sum\sum \text{cov}(X_j, X_k) \to 0$. Suppose that $E(X_j) = \mu$, $\text{cov}(X_j, X_k) = \phi_{j-k}$. Show that $\bar{X}_n \xrightarrow{\text{a.s}} \mu$ if $\sum\limits_{-\infty}^{\infty} |\phi_j| < \infty$ (an unnecessarily severe condition, but simple).

6.1.2　If $\hat{\mu}_n$ is the generic unbiased linear estimate **6.1.8**, show that

$$\text{var}(\hat{\mu}_n) = \text{var}(\mu_n^*) + \text{var}(\hat{\mu}_n - \mu_n^*),$$

which establishes the optimality of μ_n^*.

6.2　Convergence of sample averages; some stronger results

This section develops the convergence ideas of section 6.1 somewhat further. However, this is material that can well be deferred until the reading of chapter 9, in favour of the physical applications considered in the remainder of this and the next chapter.

The mean square convergence statement **6.1.6** is a special case of similar statements

$$E[H(\bar{X}_n)] \to H(\mu) \tag{6.2.1}$$

for more general functions H. It is of interest to determine for what functions, if any, **6.2.1** will hold, with a view to obtaining results which are either stronger than **6.1.6**, or which hold under weaker conditions. Furthermore, a statement such as **6.2.1** is the kind of result one needs operationally: in practice one is interested in the convergence of just such expectations.

Statement **6.2.1** implies that, at least for some functions H, the r.v. $\bar{X}_n$ behaves in the limit as though it took only the single value μ. If **6.2.1** is valid for certain H, such as $(X - \mu)^2$, this indeed implies effective equality of 'lim $\bar{X}_n$' and μ. There are, in fact, two concepts involved (convergence of r.v.s and convergence of distributions); the distinction is made in section 9.1.

It will be simpler if we work with variables $X_j - \mu$ rather than X_j (we shall assume a common expectation μ throughout). We shall continue to write the new variable as X_j, so that we can in effect set $E(X_j) = 0$ without any actual loss of generality.

Theorem 6.2.1

Suppose that

$$E(X_j | X_1, X_2, \ldots, X_{j-1}) = 0 \quad (j = 1, 2, \ldots), \tag{6.2.2}$$

and that the derivative of H obeys the uniform continuity condition

$$|H'(t+s) - H'(t)| \leq K|s|^\alpha \tag{6.2.3}$$

for some positive α. Then

$$|E[H(\bar{X}_n)] - H(0)| \leq \frac{Kn^{-\alpha-1}}{\alpha+1} \sum_{1}^{n} E(|X_j|^{\alpha+1}). \tag{6.2.4}$$

In particular,

$$E(|\overline{X}_n|^r) \leqslant \frac{4 \sum_1^n E(|X_j|^r)}{(2n)^r} \qquad \qquad 6.2.5$$

for $1 \leqslant r \leqslant 2$.

Note that we have not assumed independence of the X_j (but only the much weaker requirement **6.2.2**, cf. **5.7.7**), nor have we assumed identical distribution. A result such as **6.2.5** makes worthwhile convergence assertions possible even if variances do not exist, but only some moments of lower order.

For example, suppose we revert for explicitness to the case $E(X_j) = \mu$ for a moment; then an appeal to a Chebichev inequality and use of **6.2.5** shows that

$$P(|\overline{X}_n - \mu| > \varepsilon) \to 0 \quad \text{if} \quad n^{-r} \sum_1^n E(|X_j - \mu|^r) \to 0 \qquad 6.2.6$$

for some r in $[1, 2]$. In particular, the result will hold if the $E(|X_j - \mu|^r)$ exist, and are bounded in j, for some r greater than unity.

It is easiest to prove the theorem by considering a function $H(S_n)$ of $S_n = \sum_1^n X_j$, and thence transforming to $H(\overline{X}_n)$ by a change of scale. We have, by a partial Taylor expansion,

$$H(S_n) = H(S_{n-1}) + X_n H'(S_{n-1}) + \int_0^{X_n} [H'(S_{n-1} + s) - H'(S_{n-1})] \, ds. \qquad 6.2.7$$

Taking an expectation E_n with respect to X_n conditional on $X_1, X_2, \ldots, X_{n-1}$, we have then, in virtue of **6.2.2** and **6.2.3**,

$$|E_n H(S_n) - H(S_{n-1})]| \leqslant E_n \left(\left| \int_0^{X_n} K|s|^\alpha \, ds \right| \right) \leqslant \frac{K E_n(|X_n|^{\alpha+1})}{\alpha + 1}.$$

Continuing the procedure, we obtain

$$|E[H(S_n) - H(0)]| \leqslant \frac{K}{\alpha + 1} \sum_1^n E(|X_j|^{\alpha+1}),$$

and a change of scale $X_j \to X_j/n$ yields the result **6.2.4**. Inequality **6.2.5** follows from **6.2.4** if we note that $|t|^\alpha$ has a derivative decreasing with $|t|$ if $0 \leqslant \alpha \leqslant 1$, so that then

$$|(|t + s|^\alpha - |t|^\alpha)| \leqslant 2|\tfrac{1}{2}s|^\alpha = 2^{1-\alpha}|s|^\alpha. \qquad 6.2.8$$

Note. that this general method could still give results with weaker continuity conditions than **6.2.3**, which was chosen principally for explicitness. Actually it is not necessary to assume H differentiable at all: if H is only continuous, then we can find a function $\bar{H}$ which is arbitrarily close to it, and yet which possesses derivatives of all orders. This is achieved by the construction

$$\bar{H}(t) = E[H(t+Y)], \qquad\qquad \textbf{6.2.9}$$

where Y is a r.v. smoothly distributed in the neighbourhood of zero.

It is convenient to include here a rather stronger result than **6.2.4**, obtained assuming independence of the X_j. This is not required until section 8.6, so the reader can by-pass it for the moment.

Consider the transformation $\mathscr{F}_j H$ of a function H defined by

$$\mathscr{F}_j H(t) = E[H(t+X_j)].$$

Then $\quad \mathscr{F}_1 \mathscr{F}_2 \dots \mathscr{F}_n H(t) = E[H(t+S_n)].$

Suppose one defines the norm of a function by

$$\|H\| = \sup_t |H(t)|.$$

This norm has the properties

$$\|H+K\| \leqslant \|H\| + \|K\|, \qquad\qquad \textbf{6.2.10}$$

$$\|\mathscr{F} H\| \leqslant \|H\|. \qquad\qquad \textbf{6.2.11}$$

Consider now a second set of operators $\mathscr{G}_j$ corresponding to independent variables Y_j. We shall assume that the operators $\mathscr{F}_j$ and $\mathscr{G}_j$ can be applied independently, so it is as though the families $\{X_j\}$ and $\{Y_j\}$ were mutually independent.

Applying relations **6.2.10** and **6.2.11** in succession we find that

$$\|\mathscr{F}_1 \mathscr{F}_2 \dots \mathscr{F}_n H - \mathscr{G}_1 \mathscr{G}_2 \dots \mathscr{G}_n H\|$$

$$\leqslant \|\mathscr{F}_1 \mathscr{F}_2 \dots \mathscr{F}_n H - \mathscr{G}_1 \mathscr{G}_2 \dots \mathscr{G}_{n-1} \mathscr{F}_n H\| +$$

$$+ \|\mathscr{G}_1 \mathscr{G}_2 \dots \mathscr{G}_{n-1} \mathscr{F}_n H - \mathscr{G}_1 \mathscr{G}_2 \dots \mathscr{G}_n H\|$$

$$\leqslant \|\mathscr{F}_1 \mathscr{F}_2 \dots \mathscr{F}_{n-1} H - \mathscr{G}_1 \mathscr{G}_2 \dots \mathscr{G}_{n-1} H\| + \|\mathscr{F}_n H - \mathscr{G}_n H\|,$$

so that

$$\|\mathscr{F}_1 \mathscr{F}_2 \dots \mathscr{F}_n H - \mathscr{G}_1 \mathscr{G}_2 \dots \mathscr{G}_n H\| \leqslant \sum_1^n \|\mathscr{F}_j H - \mathscr{G}_j H\|, \qquad\qquad \textbf{6.2.12}$$

which is the key inequality we shall use later.

Exercises

6.2.1   Suppose that condition **6.2.2** holds, and H is convex. Use Jensen's inequality (exercise 2.7.9) to show that $E[H(S_n)]$ is an increasing function of n.

6.2.2   Use **6.2.12** to show that

$$\|E[H(t+\bar{X}_n)] - H(t)]\| \leqslant \sum_1^n \|E[H(t+X_j/n) - H(t)]\|$$

and hence derive **6.2.4** under the assumptions of the theorem.

6.3 Renewal processes

An economical specification of what we mean by an infinite sequence of Bernoulli trials would be to state that for such a sequence

$$E\left(\prod_1^\infty z_j^{\xi_j}\right) = \prod_1^\infty (pz_j + q),$$ **6.3.1**

where ξ_j is the r.v. which is 1 or 0 according as to whether the jth trial resulted in success or failure, and we regard **6.3.1** as valid for all $\{z_j\}$ for which the product is convergent.

In this way we can obtain the binomial distribution of the number of successes in the first n trials immediately (by setting $z_j = z$ for $j \leqslant n$, and $z_j = 1$ for $j > n$). However, we have established the geometric distribution of the number of trials to first success only by first deducing from **6.3.1** the probability of a given sequence of trial outcomes, and working on from that.

Brief and meaningful as this calculation is, one still wonders whether it is possible to deduce the result directly from **6.3.1** without the intermediate step of considering probabilities of individual sequences. In attempting to do so, we shall find that we can treat a much more general class of processes: *renewal processes* and *recurrent events*.

The practical context of a renewal process is a situation where an article, such as a machine tool, is replaced as soon as it wears out. The interest is in the probability that replacement takes place at a definite instant, and in the distribution of the number of renewals made in a given time. The situation also has a kind of converse, the idea of recurrence, which we shall consider in section 6.4. We suppose that the lifetimes of consecutive tools, denoted $T_1, T_2, \ldots$, are independent identically distributed r.v.s. Then the total lifetime of the first j tools denoted by

$$S_j = \sum_1^j T_k,$$ **6.3.2**

is just such a sum as we have considered in **5.6.6**. If the first article (the one of lifetime T_1) was installed at time zero, then S_j is also the instant at which the jth renewal takes place.

We shall assume that lifetimes T_k are integral valued, so that we can work in integral time, n. The assumption makes for simplicity without losing much realism: we are in effect rounding off lifetimes to the nearest whole number of time units. The r.v. 'lifetime' will then have a p.g.f.

$$\Pi(z) = \sum_0^\infty p_t z^t,$$

where p_t is the probability of a lifetime $T = t$. It is usual to set $p_0 = 0$, that is, to exclude the possibility that an article has zero lifetime, and needs replacement the very moment it is installed. We shall not require this, but merely demand that $p_0 < 1$, that is, that the article does not *always* fail immediately!

The p.g.f. $\Pi(z)$ is always convergent in $|z| \leqslant 1$, and, because of the assumption $p_0 < 1$, we shall have

$$|\Pi(z)| < 1 \quad \text{if} \quad |z| < 1.$$

Let R_n be the number of renewals made *at* time n, and let

$$u_n = E(R_n). \tag{6.3.3}$$

The installation of the original article is counted in R_0. If $p_0 = 0$, then R_n can only take the values 0 or 1, and u_n then has the interpretation of the *probability of renewal at time n.*

Theorem 6.3.1

If $|z| < 1$, then

$$\sum_0^\infty R_n z^n = \sum_{j=0}^\infty z^{S_j} \tag{6.3.4}$$

(with the convention $S_0 = 0$), and

$$\sum_0^\infty u_n z^n = \frac{1}{1 - \Pi(z)}. \tag{6.3.5}$$

Relation **6.3.4** is the key identity. Its formal validity is evident, since a renewal at time n will contribute z^n to each side : the question is whether these infinite series with random coefficients converge. If $p_0 = 0$ then R_n can only take the values 0 or 1, so the series are dominated by $\sum |z|^n$, and necessarily converge. In the case $0 < p_0 < 1$ a slightly more careful argument is needed, which we shall defer until section 9.3. This yields the conclusion that the series converge in the conventional sense, with probability one. (This is a concept of-convergence differing from those used in section 6.1 : it is known as *almost certain convergence.*)

The second result **6.3.5** follows from **6.3.4** : if the generating function of the u_n is denoted $U(z)$, then by taking expectations of **6.3.4** and appealing to the property **5.6.8** we have

$$U(z) = \sum_{j=0}^\infty \Pi(z)^j = \frac{1}{1 - \Pi(z)}. \tag{6.3.6}$$

This essentially determines the u_n in terms of the lifetime distribution.

Let us consider a pair of simple examples. Suppose that lifetime is fixed, and equal to τ, so that $\Pi(z) = z^\tau$.

Then $\quad U(z) = \dfrac{1}{1 - z^\tau} = \sum_{k=0}^\infty z^{k\tau},$

with the obvious interpretation that there is a single renewal when n equals a multiple of τ, and at no other time.

If we consider a geometric lifetime distribution, so that

$$\Pi(z) = \frac{pz}{1-qz},$$ 6.3.7

then **6.3.5** yields

$$U(z) = 1 + \frac{pz}{1-z},$$ 6.3.8

or $u_0 = 1$; $u_n = p$ $(n > 0)$. That is, the renewal probability is constant after the initial installation; a fact whose significance will be brought out in section 6.4.

Exercises

6.3.1 Calculate p_t and u_n for the cases $\Pi(z) = \left(\dfrac{p}{1-qz}\right)^m$ $(m = 1, 2)$. What is the limit value of u_n for large n?

6.3.2 Show that $u_n \leqslant 1/(1-p_0)$.

6.3.3 Relation **6.3.5** is equivalent to the equations

$$u_0 = 1 + p_0 u_0,$$

$$u_n = \sum_{t=0}^{n} p_t u_{n-t} \quad (n = 1, 2, \ldots).$$

Deduce these directly.

6.3.4 Interpret the coefficient of $s^j z^n$ in the expansion of $[1 - s\,\Pi(z)]^{-1}$ in non-negative powers of s and z.

6.3.5 Show from **6.3.4** that

$$\sum_{0}^{\infty}\sum_{0}^{\infty} z^j w^k \operatorname{cov}(R_j, R_k) = \frac{\Pi(zw) - \Pi(z)\Pi(w)}{[1 - \Pi(z)][1 - \Pi(w)][1 - \Pi(zw)]}.$$

A r.v. of interest is

$$N_n = \sum_{0}^{n} R_k,$$ 6.3.9

the *total number of renewals* made up to time n. For this variable, the key fact is that $N_n < j$ is the same event as $S_j > n$; essentially the fact embodied in **6.3.4**. One can use this relation to obtain exact results: we shall use it to obtain the approximate relations

$$E(N_n) = \frac{n}{\mu} + o(n),$$ 6.3.10

$$\operatorname{var}(N_n) = \frac{n\sigma^2}{\mu^3} + o(n),$$ 6.3.11

where μ and σ^2 are respectively the mean and the variance of lifetime.

For simplicity of notation we shall denote N_n simply by N. Let A_j denote the event $N < j$. Then note that

$$0 \leqslant E(S_j | \bar{A}_j) \leqslant n, \tag{6.3.12}$$

and that if $j \geqslant n$, then

$$P(\bar{A}_j) \leqslant p_0^{j-n}, \tag{6.3.13}$$

since there must then be at least $j - n$ zero lifetimes.

We have now

$$j\mu = E(S_j) = P(A_j)E(S_j | A_j) + P(\bar{A}_j)E(S_j | \bar{A}_j). \tag{6.3.14}$$

$$\text{But} \quad E(S_j | A_j) = E(S_j - S_N + S_N | A_j) = E[\mu(j - N) + S_N | A_j], \tag{6.3.15}$$

since S_N is the instant of first renewal after time n, and, conditional on the value of N, $S_j - S_N$ is freely distributed as the sum of $j - N$ unconditioned lifetimes. From **6.3.14** and **6.3.15** we have thus

$$P(A_j)E(S_N - \mu N | A_j) + P(\bar{A}_j)[E(S_j | \bar{A}_j) - j\mu] = 0. \tag{6.3.16}$$

Letting j tend to infinity in **6.3.16** we obtain, in virtue of **6.3.12** and **6.3.13**,

$$E(S_N - \mu N) = 0. \tag{6.3.17}$$

The expectation conditional on A_j in **6.3.16** tends to an unconditioned one, since by **6.3.13** A_j tends to the certain event. We see from **6.3.17** that

$$E(N) = \frac{E(S_N)}{\mu}. \tag{6.3.18}$$

Now, S_N will be greater than n, but only by an amount of the order of a single lifetime. If we make the very common approximation of 'neglecting overshoot', we deduce **6.3.10** from the exact result **6.3.18**.

By applying the argument starting at **6.3.14** to $E(S_j^2)$ we obtain, analogously to **6.3.17**,

$$E[(S_N - \mu N)^2 - \sigma^2 N] = 0, \tag{6.3.19}$$

which leads to **6.3.11**.

Since $E(N_n) = \sum_0^n u_k$ we see that **6.3.10** implies that

$$\frac{1}{n}\sum_0^n u_k \to \frac{1}{\mu} \tag{6.3.20}$$

with increasing n. Under wide conditions one can prove the much stronger result,

$$u_n \to \frac{1}{\mu}, \tag{6.3.21}$$

that the expected number of renewals each unit time tends to a constant, which must necessarily be the reciprocal of expected lifetime. A general proof is not straightforward, although the result is fairly immediate in simple cases (see exercise 6.3.8).

Exercises

6.3.6   Confirm **6.3.21** for the cases of exercise 6.3.1.

6.3.7   Show that the equation $\Pi(z) = 1$ has no roots inside the unit circle. Show also that if we require that lifetime distribution be *aperiodic* (i.e. we exclude the case where lifetime distribution is concentrated on multiples of some integer greater than unity) then the only root on the unit circle is a simple one at $z = 1$.

6.3.8   Consider the case when lifetime is bounded (so that $\Pi(z)$ is a polynomial) and the lifetime distribution is aperiodic. Show from the partial fraction expansion of $[1 - \Pi(z)]^{-1}$ that

$$u_n = \mu^{-1} + O(z_0^{-n}),$$

where z_0 is the root of $\Pi(z) = 1$ smallest in modulus after $z = 1$.

6.3.9   *Wald's identity.* By applying the argument beginning at **6.3.14** to $E(z^{S_j})$, show that

$$E(z^{S_N}\Pi(z)^{-N}) = 1 \qquad\qquad\qquad \textbf{6.3.22}$$

for all z such that $\Pi(z)$ exists and $|\Pi(z)| > p_0$.

Note that **6.3.17** and **6.3.19** follow from differentiation of the identity with respect to z at $z = 1$. There are two r.v.s in the bracket of **6.3.22**: N and S_N. The 'no overshoot' approximation is to set $S_N \simeq n$.

6.4 Recurrent states (events)

There are cases when it is the u_n rather than the lifetime probabilities p_t which are known, so that in this case one would invert relation **6.3.6** to obtain

$$\Pi(z) = \frac{U(z) - 1}{U(z)}. \qquad\qquad\qquad \textbf{6.4.1}$$

For example, this is strictly the situation when one is considering occurrence of successes in a sequence of Bernoulli trials. One knows from **6.3.1** that there is a constant probability of success p, so that $u_n = p$ $(n > 0)$, with $u_0 = 1$, by convention. The generating function is thus given by **6.3.8**, and substitution into **6.4.1** gives the p.g.f. $\Pi(z)$ of **6.3.7**. Thus we deduce that the 'lifetime distribution', that is distribution of number of trials between consecutive successes, is geometric.

Of course, for this converse argument to hold it is necessary that the basic hypothesis of the renewal process be fulfilled: that the intervals between consecutive occurrences of a success be independent, identically distributed variables. This is evident in the case of a Bernoulli sequence.

In general, suppose that a situation $\mathscr{A}$ can occur from time to time in a sequence of trials (not necessarily independent) and that the process is such that the numbers of trials between consecutive occurrences of $\mathscr{A}$ are independent, identically distributed r.v.s. Then $\mathscr{A}$ is termed a *recurrent event* or *recurrent state*. The second term is better, since the first conflicts with our previous use of the technical term 'event'. We shall clarify the point and formalize the definition after examining a few more examples.

Thus, 'renewal' was a recurrent state in the last section; 'success' is a recurrent state for Bernoulli sequence. In many cases the u_n are quickly determinable, and are interpretable as $P(\mathscr{A}$ occurs at nth trial), because multiple occurrences are excluded (i.e. $p_0 = 0$). Equation **6.4.1** then determines the p.g.f. $\Pi(z)$ of *recurrence times* of $\mathscr{A}$. In effect, it determines the probability that $\mathscr{A}$ *first occurs* at n (after the occurrence at 0) in terms of the probability that $\mathscr{A}$ *occurs* at n.

For example, consider again a sequence of Bernoulli trials, and let $\mathscr{A}$ be the situation that the number of successes equals the number of failures – this certainly occurs at $n = 0$, and is a recurrent state. If one were playing a series of games for constant stakes, the times when $\mathscr{A}$ occurred would be just those at which one broke even.

Since u_n is the probability of $\frac{1}{2}n$ successes in n trials, we have, by the binomial distribution,

$$
u_n = \begin{cases} 0 & (n \text{ odd}), \\ \binom{2r}{r}(pq)^r & (n = 2r). \end{cases}
$$

6.4.2

The series $U(z)$ can be summed; we have

$$
\begin{aligned}
U(z) &= \sum_{r=0}^{\infty} \binom{2r}{r}(pq)^r z^{2r} \\
&= (1 - 4pqz^2)^{-\frac{1}{2}},
\end{aligned}
$$

6.4.3

as the reader can verify directly (see also exercise 6.4.4); that root being taken which tends to unity as z tends to zero. Thus the recurrence time p.g.f. is, by **6.4.1**,

$$
\Pi(z) = 1 - (1 - 4pqz^2)^{\frac{1}{2}},
$$

6.4.4

so that a recurrence time t has probability

$$
p_t = \begin{cases} 0 & (t \text{ odd or zero}), \\ \binom{2r}{r}\dfrac{(pq)^r}{2r-1} & (t = 2r; r = 1, 2, \ldots). \end{cases}
$$

6.4.5

Note that the distribution is not an aperiodic one, but is restricted to even values of t, as it obviously must be.

A very interesting point is that

$$\Pi(1) = 1 - (1 - 4pq)^{\frac{1}{2}}$$

$$= 1 - |p - q|, \qquad \qquad \textbf{6.4.6}$$

so that $\Pi(1) < 1$ unless $p = q$, that is, unless the game is a fair one. Now, $\Pi(1) = \sum p_t$, and for all distributions we have ever encountered, $\sum p_t = 1$. How are we then to interpret **6.4.6**?

The actual interpretation of the infinite sum $\sum p_t$ is as a limit such as $\lim_{\tau \to \infty} \sum_0^{\tau} p_t$ or $\lim_{z \uparrow 1} \sum_0^{\infty} p_t z^t$. The first interpretation makes it plain that the sum is to be regarded as *the probability that recurrence time T is finite*. The inequalities

$$\sum_0^{\tau} p_t \leqslant \lim_{z \uparrow 1} \sum_0^{\infty} p_t z^t \leqslant \lim_{\tau \to \infty} \sum_0^{\tau} p_t \qquad \qquad \textbf{6.4.7}$$

show that these two limits must be equal, so that $\Pi(1)$ can also be interpreted as the probability of a finite recurrence time. All the r.v.s considered hitherto have been finite with probability one, and here we have the first exception, and in a natural problem. If there is a deficit, $\Pi(1) < 1$, we shall say that recurrence is *uncertain*. The reason for uncertain recurrence is intuitively clear in the present case. If the game is an unfair one, the player with an advantage has a good chance of building up a substantial lead: this, once achieved, is likely never to be lost.

However, even in the case $p = q = \frac{1}{2}$, when the game is a fair one, the time to recurrence is extraordinarily long. The mean recurrence time, if it exists, should be given by $\Pi'(1)$; we see from **6.4.4** that this is infinite. In fact, an application of Stirling's approximation to expression **6.4.5** shows that

$$p_t \simeq \sqrt{\frac{2(4pq)^t}{\pi t^3}}, \qquad \qquad \textbf{6.4.8}$$

for t large and even. Thus if $p = q$ the distribution tails away at the slow rate of $t^{-\frac{3}{2}}$, implying that large recurrence times are relatively probable.

The objection to the term 'recurrent event' is that an event is something which for a given experimental outcome occurs either once or not at all (the 'outcome' in the present case being the composite observation constituted by the *whole sequence* of trials) – it cannot happen repeatedly in the sequence.

Suppose that ξ_n describes the outcome of the nth trial, and that we consider a sequence of r.v.s

$$\eta_n = \eta(\xi_n, \xi_{n-1}, \ldots), \qquad \qquad \textbf{6.4.9}$$

where η is a function which can only take the values 0 or 1. Let 'occurrence of $\mathscr{A}$ at n' correspond to $\eta_n = 1$: this is a true event, which could be denoted by

A_n. Suppose that $\eta_n = 1$ for $n = 0, n_1, n_2, \ldots$. If $n_1, n_2 - n_1, n_3 - n_2, \ldots$ are independent identically distributed r.v.s, then the state $\mathscr{A}$ is recurrent, by definition.

Thus for the example of this section, $\eta_n = 1$ if $S_n = \sum\limits_{}^{n} \xi_k = 0$, where ξ_k takes values $+1$ or -1 for success or failure at the kth trial; if $S_n \neq 0$ then $\eta_n = 0$.

To establish whether a given $\mathscr{A}$ is recurrent or not is not always a trivial matter. One pair of sufficient conditions that $\mathscr{A}$ be recurrent is that

$$E(Y_n \mid A_n; \xi_n, \xi_{n-1}, \ldots) = E(Y_n \mid A_n), \qquad \textbf{6.4.10}$$

where Y_n is any r.v. defined on the 'future' at n, (i.e. a function only of $\xi_{n+1}, \xi_{n+2}, \ldots$); and that expression **6.4.10** be independent of n if Y_n is formed in a fixed way. What **6.4.10** states is that, if $\mathscr{A}$ is known to have occurred at n, then the future of the process is affected only by this fact, and the additional conditioning provided by knowledge of $\xi_n, \xi_{n-1}, \ldots$ has no effect. Occurrence of $\mathscr{A}$ constitutes what is called a *regeneration point* of the process; in effect the process then makes a fresh start.

So, for the gaming problem of this section, if we know that $S_n = 0$, then the process starts again in exactly the same conditions as those we began with at $n = 0$. It matters only that $S_n = 0$; additional specification of $\xi_n, \xi_{n-1}, \ldots$ would not affect the outcomes of future trials.

Exercises

6.4.1 Show for case **6.4.2** that, for recurrence time T,

$$E(T \mid T < \infty) = 1 + \frac{1}{|p - q|}.$$

6.4.2 Show for case **6.4.2** that $u_n \to 0$ with increasing n, which, in the case of certain recurrence ($p = q$), is consistent with **6.3.21** and $\mu = \infty$.

6.4.3 Note that $U(1) - 1$ is the expected number of recurrences, and that $\Pi(1) < 1$ corresponds to $U(1) < \infty$.

6.4.4 Show that the infinite sum in **6.4.3** is the absolute term in the Laurent expansion of $[1 - \frac{1}{2}z(pw + qw^{-1})]^{-1}$ in powers of w on the unit circle, and hence derive the final expression in **6.4.3**.

6.4.5 Let the number of successes less the number of failures in a Bernoulli sequence be denoted S_n. Is $S_n = k$ a recurrent state? Is $S_n > k$ a recurrent state?

6.4.6 Suppose that $\mathscr{A}$ and $\mathscr{B}$ are states whose occurrence constitutes a regeneration point. Let

$$U_{\mathscr{A}\mathscr{B}}(z) = \sum_{0}^{\infty} z^n P(\mathscr{B} \text{ at } n \mid \mathscr{A} \text{ at } 0),$$

$$U_{\mathscr{B}\mathscr{B}}(z) = \sum_{0}^{\infty} z^n P(\mathscr{B} \text{ at } n \mid \mathscr{B} \text{ at } 0),$$

and let $\Pi_{\mathscr{B}\mathscr{B}}(z)$ and $\Pi_{\mathscr{A}\mathscr{B}}(z)$ be correspondingly the p.g.f. of recurrence time to $\mathscr{B}$, and of *first passage time* from $\mathscr{A}$ to $\mathscr{B}$, so that $U_{\mathscr{B}\mathscr{B}} = (1 - \Pi_{\mathscr{B}\mathscr{B}})^{-1}$. Show that $U_{\mathscr{A}\mathscr{B}} = \Pi_{\mathscr{A}\mathscr{B}} U_{\mathscr{B}\mathscr{B}}$.

6.4.7   Suppose that the states $\mathscr{A}$ and $\mathscr{B}$ correspond to $S = 0$ and $S = b$ in a Bernoulli process (cf. exercise 6.4.5). Show (cf. exercise 6.4.4) that

$$U_{\mathscr{A}\mathscr{B}} = \frac{\{(1 - \Delta)/2qz\}^b}{\Delta} = \frac{\phi^b}{\Delta},$$

say, where $\Delta = \sqrt{(1 - 4pqz^2)}$, and hence that $\Pi_{\mathscr{A}\mathscr{B}} = \phi^b$. Hence show that the probability that S ever equals b is 1 if $p \geqslant q$, and $(p/q)^b$ otherwise.

6.5   A result in statistical mechanics

A model for a system such as a perfect gas would be somewhat as follows. Imagine a collection of n molecules, each possessing a certain amount of energy; these individual energies being restricted to a set of possible values $\varepsilon_1, \varepsilon_2, \ldots$. It is assumed that there is no energy of interaction, so that the total energy of the system is

$$\mathscr{E} = \sum_j n_j \varepsilon_j, \tag{6.5.1}$$

where n_j is the number of molecules at energy level ε_j.

Suppose that the total energy $\mathscr{E}$ is prescribed, as is the total number of molecules

$$n = \sum_j n_j. \tag{6.5.2}$$

Then a basic theorem in statistical mechanics states that all possible allocations of the energy $\mathscr{E}$ among the n (distinguishable) molecules are equally likely. This principle is subject to much qualification, modification, and refinement, but for present purposes we shall take it in the simple form just stated. Thus, any allocation $(n_1, n_2, \ldots)$ which is consistent with 6.5.1 and 6.5.2 has probability proportional to

$$\frac{n!}{\prod n_j!}. \tag{6.5.3}$$

Suppose that we now let n and $\mathscr{E}$ tend to infinity in constant ratio, rather as we did with n and V to derive the Poisson process of section 4.7. That is, we consider an infinite 'sea' of molecules with an average individual energy $\bar{\varepsilon} = \lim(\mathscr{E}/n)$. Then, a second basic theorem of statistical mechanics, derived from the one just stated, asserts that the proportion of molecules in the jth energy level, n_j/n, tends to

$$\pi_j = e^{-\lambda - \mu \varepsilon_j}, \tag{6.5.4}$$

where λ and μ are a pair of constants determined by

$$\sum \pi_j = 1,$$ 6.5.5

$$\sum \pi_j \varepsilon_j = \bar{\varepsilon}.$$ 6.5.6

The usual 'proof' is as follows. The logarithm of expression 6.5.3 is

$$L \simeq \text{constant} + \sum n_j - \sum n_j \log n_j,$$ 6.5.7

if we take the leading terms in Stirling's approximation for $\log(n_j!)$. Now, in order to find the 'most probable' values of the n_j, maximize L with respect to the n_j, subject to constraints 6.5.5 and 6.5.2. If we introduce Lagrangian multipliers λ' and μ for these constraints, then we have the task of maximizing $L - \lambda' \sum n_j - \mu \sum n_j \varepsilon_j$. If we neglect the fact that the n_j must be integer-valued, then we find that the maximum is attained for

$$n_j \simeq \bar{n}_j = e^{-\lambda' - \mu \varepsilon_j}.$$ 6.5.8

If one assumes that the maximum becomes indefinitely sharp with increasing n, i.e. that $P\left(\left| \dfrac{n_j - \bar{n}_j}{n} \right| > \delta \right) \to 0$, then 6.5.8 implies 6.5.4.

This proof leaves many gaps, the obvious one being that last mentioned: whether by locating the most probable values of n_j/n we have indeed located true 'stochastic limit values'. However, the neglect of the integral character of the n_j raises an even more delicate point: for given $\mathscr{E}$ and ε_j, do 6.5.1 and 6.5.2 possess sufficiently many solutions in integers n_j that these approximate methods work at all? Suppose, for example, that $\varepsilon_j = \log p_j$, where p_j is the jth prime. Then, since the decomposition of an integer into prime factors

$$e^{\mathscr{E}} = \prod_j p_j^{n_j}$$

is unique, there is at most one set of n_j, say n_j', satisfying 6.5.1 and 6.5.2. The proportion of molecules at level $\mathscr{E}_j$ is thus *fixed* at n_j'/n, and there is no reason at all why these ratios should be consistent with 6.5.4.

One way out of the difficulty is to make the requirement 6.5.1 less rigid; and to replace it by something like

$$\mathscr{E} - \Delta\mathscr{E} \leqslant \sum n_j \varepsilon_j \leqslant \mathscr{E} + \Delta\mathscr{E}.$$ 6.5.9

That is, we let energy take values in a band, which, although relatively narrow, allows the n_j enough freedom that the argument leading to 6.5.4 can be made valid.

However, we shall consider a particular case which can be solved exactly, and for which there is a complete and simple treatment, verifying 6.5.4. For us, it is an example on the multinomial distribution 6.5.3 with a conditioning relation 6.5.1.

The case is that in which the energy levels constitute an arithmetic progression:

$$\varepsilon_j = j\varepsilon \quad (j = 0, 1, 2, \ldots).$$ 6.5.10

We must then assume that the prescribed total energy $\mathscr{E}$ is also a multiple, $m\varepsilon$, of ε.

Suppose that there were no energy constraint 6.5.1, so the distribution of the occupation numbers would be simply multinomial. The total energy of any configuration would be a r.v., $M\varepsilon$, say. The joint p.g.f. of M and the occupation numbers n_j would be

$$\Pi(z, w) = E\left(w^M \prod_j z_j^{n_j}\right) = \left(\frac{\sum z_j w^j}{n}\right)^n . \qquad \textbf{6.5.11}$$

Applying the energy constraint, we find then that the p.g.f. of the n_j conditional on $M = m$ is

$$\Pi(z) = \frac{\text{coefficient of } w^m \text{ in } \left(\sum z_j w^j\right)^n}{\text{coefficient of } w^m \text{ in } \left(\sum w^j\right)^n}, \qquad \textbf{6.5.12}$$

cf. 5.1.4.

In particular, the conditional expectation of n_j is

$$E(n_j | M = m) = \left[\frac{\partial \Pi(z)}{\partial z_j}\right]_{z=1}$$

$$= \frac{\text{coefficient of } w^m \text{ in } n w^j \left(\sum w^j\right)^{n-1}}{\text{coefficient of } w^m \text{ in } \left(\sum w^j\right)^n}$$

$$= n \frac{\text{coefficient of } w^{m-j} \text{ in } (1-w)^{-n+1}}{\text{coefficient of } w^m \text{ in } (1-w)^{-n}}$$

$$= n(n-1) \frac{m^{(j)}}{(m+n-1)^{(j+1)}}. \qquad \textbf{6.5.13}$$

This is an exact result. Now let m and n tend to infinity in such a way that $m/n \to \alpha$. Then

$$E\left[\frac{n_j}{n}\middle| \text{energy per molecule} = \alpha\varepsilon\right] \to \frac{1}{1+\alpha}\left(\frac{\alpha}{1+\alpha}\right)^j \quad (j = 0, 1, 2, \ldots), \qquad \textbf{6.5.14}$$

which agrees with 6.5.4 and 6.5.10. Moreover, it is an elementary matter to show that n_j/n converges in mean square to expression 6.5.14.

Exercises

6.5.1   Show that, for case **6.5.10**, $\text{cov}(n_j/n, n_k/n) = O(n^{-1})$ if m and n are of the same order, and j, k are held fixed. Hence show that n_j/n tends to expression **6.5.14** in mean square.

6.5.2   Show that the general result corresponding to **6.5.13** is

$$E\left(\frac{n_j}{n}\,\middle|\,\sum n_k \varepsilon_k = \mathscr{E}\right) = \frac{\text{coefficient of } e^{\theta(\mathscr{E}-\varepsilon_j)} \text{ in } \left(\sum e^{\theta \varepsilon_k}\right)^{n-1}}{\text{coefficient of } e^{\theta \mathscr{E}} \text{ in } \left(\sum e^{\theta \varepsilon_k}\right)^n}.$$

6.5.3   Note that we could let the number of particles vary, as well as the energy, and so characterize $\Pi(z)$ as being proportional to the coefficient of $\zeta^n e^{\theta \mathscr{E}}$ in the expansion of

$$\phi(z, \zeta, \theta) = \prod_{j \geq 0}^{\infty} \exp(\zeta z_j e^{\theta \varepsilon_j}).$$

This is for *Boltzmann* statistics; for *Bose–Einstein* statistics (when molecules are regarded as indistinguishable) and *Fermi–Dirac* statistics (when at most one molecule can occupy a given energy level) we have

$$\phi = \prod_{0}^{\infty} (1 \pm \zeta z_j e^{\theta \varepsilon_j})^{\pm 1}$$

if we take the lower and upper option of sign respectively. Note the similarity of the Fermi–Dirac case to the problem of exercise 4.5.5: that of determining the distribution of total income in a sample of n chosen without replacement.

6.6   Branching processes

In 1873–4, Galton and de Candolle remarked on the many instances of family names that had become extinct, and, prompted by this, raised the general question: what is the probability that a natural population dies out in the course of time? Such a question cannot be answered by appeal to a deterministic model: the model must have a probabilistic element. For example, suppose that we can measure time in generations, and that X_n is the number of population members in the nth generation. Then the simplest deterministic model would be

$$X_{n+1} = \alpha X_n, \qquad\qquad\qquad\qquad **6.6.1**$$

where α is the multiplication factor from one generation to the next. The population will increase indefinitely, tend to zero, or just maintain itself, according as to whether α is greater than, less than, or equal to unity. But these statements are too crude; particularly in small populations. They take no account of fluctuations, or even of the fact that X_n must be integer-valued. One would like to be able to determine, for example, the probability that the line descending from a single initial ancestor ultimately becomes extinct.

A model for this situation was proposed and partly analysed by a clergyman, H. W. Watson, in 1874; the analysis was completed by J. F. Steffensen in 1930. The model is the simplest example of what we now know as a *branching process*. It serves as a model, not only for population growth, but for other multiplicative phenomena, such as the spread of an infection or the progress of a nuclear fission reaction. The event 'indefinite survival of a population' corresponds in these two cases to the occurrence of an epidemic, or of a nuclear explosion, respectively.

For concreteness we shall formulate the model in terms of the surname example and make the following idealizing assumptions: that the numbers of sons of different individuals (in whatever generation) are independent and identically distributed r.v.s. We restrict attention to male members of a line, because it is only through these that the family name is inherited. Thus, the probability that a man has j sons is, say, $p_j (j = 0, 1, 2, \ldots)$, independently of the number of individuals in his or previous generations, and of the numbers of sons sired by other members of his generation. We can thus define a *progeny p.g.f.*

$$G(z) = \sum_0^\infty p_j z^j, \qquad\qquad\qquad 6.6.2$$

with the property

$$E(z^{X_{n+1}} \mid X_n = 1) = G(z), \qquad\qquad\qquad 6.6.3$$

where X_n is the number in the nth generation.

The model is thus a very idealized one: descendants are rather arbitrarily grouped into generations (which may ultimately overlap in time); effects such as environmental limitation or variation of birth rate with population size are neglected; the numbers of males only are considered. However, the model is still a valuable generalization of **6.6.1**, and produces some interesting new effects and ideas.

If $X_n = k$, then X_{n+1} is the sum of k independent r.v.s each with p.g.f. $G(z)$, so that relation **6.6.3** generalizes to

$$E(z^{X_{n+1}} \mid X_n) = G(z)^{X_n}. \qquad\qquad\qquad 6.6.4$$

Let $\quad \Pi_n(z) = E(z^{X_n}) \qquad\qquad\qquad 6.6.5$

be the p.g.f. of X_n. Then, averaging relation **6.6.4** over the values of X_n, we find that

$$\Pi_{n+1}(z) = \Pi_n[G(z)]. \qquad\qquad\qquad 6.6.6$$

This is the basic relation between the p.g.f.s $\Pi_n(z)$. It is a generalization of the recursion **6.6.1**, and is, of course, very much more informative, since it relates distributions rather than simple numbers.

Let us formalize our conclusion.

Theorem 6.6.1

For the simple branching process the p.g.f. $\Pi_n(z)$ *of the number in the n-th genera-*
tion obeys recursion **6.6.6**, *where* $G(z)$ *is the progeny p.g.f. If* $X_0 = 1$, *so that*
$\Pi_0(z) = z$, *then*

$$\Pi_n(z) = G^{(n)}(z), \qquad\qquad\qquad 6.6.7$$

where $G^{(n)}(z)$ *is the n-th iterate of* $G(z)$.

The second statement, for the case of a single ancestor in the 'zeroth'
generation, follows from **6.6.6**. We have $\Pi_1(z) = G(z)$, $\Pi_2(z) = G[G(z)]$, and
in general **6.6.7** holds, where $G^{(n)}(z)$ is the function obtained by applying the
transformation $z \to G(z)$ to z n times.

Equation **6.6.7** solves the problem in the same sense that the relation

$$X_n = \alpha^n \qquad\qquad\qquad 6.6.8$$

solves the model **6.6.1** : it determines the distribution of X_n as explicitly as is
generally possible. The probabilistic problem is thus reduced to the analytic
one of calculating the nth iterate of a function $G(z)$. This problem is classic and
difficult, and one can solve it explicitly only in a few cases (see exercises 6.6.4
and 6.6.5). However, one can extract a certain amount of useful information
from relations **6.6.6** and **6.6.7** without actually evaluating $G^{(n)}(z)$.

Exercises

6.6.1   Show that the assumptions made imply the stronger form of equation **6.6.4** :

$$E(z^{X_{n+1}} \mid X_0, X_1, X_2, \ldots, X_n) = G(z)^{X_n}.$$

6.6.2   Show that the general solution of equation **6.6.6** is

$$\Pi_n(z) = \Pi_0[G^{(n)}(z)].$$

6.6.3   Consider the case of one initial ancestor, when

$$\Pi_n(z) = E(z^{X_n} \mid X_0 = 1).$$

Show from first principles, without appealing to the solution **6.6.7**, that relation
6.6.6 can be generalized to

$$\Pi_{m+n}(z) = \Pi_m[\Pi_n(z)] \quad (m, n = 0, 1, 2, \ldots).$$

6.6.4   Suppose that

$$G(z) = \frac{pz+q}{1+\gamma-\gamma z}. \qquad\qquad\qquad 6.6.9$$

Show that this is the p.g.f. of a non-negative r.v. if $p+q = 1$; $p, q, \gamma \geqslant 0$. Show
also that

$$G^{(n)}(z) = \frac{A_n z + B_n}{C_n z + D_n},$$

where

$$\begin{bmatrix} A_n & B_n \\ C_n & D_n \end{bmatrix} = \begin{bmatrix} p & -\gamma \\ q & 1+\gamma \end{bmatrix}^n.$$

6.6.5 Show that

$$G(z) = 1 - \gamma(1-z)^\delta$$

is the p.g.f. of a non-negative r.v. if $0 \leqslant \gamma, \delta \leqslant 1$, and evaluate its nth iterate.

6.6.6 Show that if $G(z)$ is a p.g.f. of a non-negative r.v., and can be represented as $G(z) = J^{-1}[1+J(z)]$, then $G^{(n)}(z) = J^{-1}[n+J(z)]$, and this is also a p.g.f.

We can use the recursion **6.6.6** to determine the moments of X_n in a relatively straightforward fashion. Differentiating once, and setting $z = 1$, we obtain

$$E(X_{n+1}) = \alpha E(X_n), \qquad\qquad\qquad\qquad \textbf{6.6.10}$$

where $\quad \alpha = G'(1) = \sum_0^\infty j p_j, \qquad\qquad\qquad\qquad \textbf{6.6.11}$

the expected number of sons born to a man.

This relation corresponds nicely to the deterministic recursion **6.6.1**, and has the general solution

$$E(X_n) = \alpha^n E(X_0). \qquad\qquad\qquad\qquad \textbf{6.6.12}$$

Differentiating **6.6.6** twice at $z = 1$, we obtain, with some reduction,

$$\mathrm{var}(X_{n+1}) = \alpha^2 \, \mathrm{var}(X_n) + \beta E(X_n), \qquad\qquad\qquad\qquad \textbf{6.6.13}$$

where β is the variance of the number of sons born to a man. This difference equation has the solution, in terms of values at $n = 0$,

$$\mathrm{var}(X_n) = \alpha^{2n} \, \mathrm{var}(X_0) + \frac{\beta \alpha^{n-1}(\alpha^n - 1)}{\alpha - 1} E(X_0). \qquad\qquad \textbf{6.6.14}$$

One can continue in this way and calculate the moments of X_n as far as one has patience. However, it is more illuminating to calculate the *extinction probability*, of which we spoke at the very beginning of the section. Suppose that

$$\rho_n = P(X_n = 0) = \Pi_n(0), \qquad\qquad\qquad\qquad \textbf{6.6.15}$$

so that ρ_n is the probability of extinction by the nth generation. This is nòt to be confused with the event 'extinction *at* the nth generation', which would have probability $\rho_n - \rho_{n-1}$. The sequence $\{\rho_n\}$ is non-decreasing (because of this last remark), and since it is also bounded, it must then have a limit value

$$\rho = \lim_{n \to \infty} \rho_n. \qquad\qquad\qquad\qquad \textbf{6.6.16}$$

We shall interpret ρ as 'the probability of extinction in a finite time', or as 'the probability of ultimate extinction'.

Suppose we consider the case of a single initial ancestor, so that **6.6.7** holds. Then, from **6.6.7** and **6.6.15**,

$$\rho_{n+1} = G^{(n+1)}(0) = G[G^{(n)}(0)] = G(\rho_n) \quad (n = 0, 1, 2, \ldots) \tag{6.6.17}$$

and this recursion, together with the initial condition $\rho_0 = 0$, determines the sequence $\{\rho_n\}$. By letting n tend to infinity in **6.6.17** we see that the limiting extinction probability ρ must be a root of the equation

$$G(z) = z. \tag{6.6.18}$$

This interesting result relates the extinction probability directly tó the progeny p.g.f. By using arguments which are basically quite simple, we can complete the result to obtain the following conclusions:

Theorem 6.6.2

*If $G(z)$ is not identically equal to z (i.e. if a man does not have exactly one son with probability one) then equation **6.6.18** has just two real positive roots, of which $z = 1$ is always one. The extinction probability ρ is the smaller of the two roots; also $\rho < 1$ if $\alpha > 1$ and $\rho = 1$ if $\alpha \leqslant 1$.*

This last result tallies with one's conclusions from the deterministic case, although it is interesting to note that extinction is certain also in the transitional case $\alpha = 1$, when a man replaces himself exactly on average.

We shall give an outline proof of the result, and take up the neglected details in exercise 6.6.13.

Note first that, since $G(z)$ is a power series with positive coefficients, its derivatives exist in any open real interval for which $G(z)$ converges (certainly for $0 \leqslant z < 1$) and are themselves positive. In particular, since $G(z)$ has an increasing first derivative (i.e. it is *convex*) the graph of $G(z)$ is intersected by any straight line in at most two points. Thus equation **6.6.18** has at most two roots and one of these is certainly $z = 1$.

We leave it to the reader to convince himself that the cobweb construction in Figure 6 generates the sequence $\{\rho_n\}$ of **6.6.17**. Graphically, it is obvious that this sequence converges to the first intersection of the two curves, i.e. to the smaller positive root of **6.6.18**. The point is proved analytically in exercise 6.6.13.

Now, since at the smaller root the graph of $G(z)$ crosses the 45° line from above, we must have $G'(\rho) < 1$, if the two roots of **6.6.18** (say ρ and ρ') are distinct, and $G'(\rho') > 1$. Now, since one of ρ and ρ' is 1, and $\alpha = G'(1)$, we must have $\rho = 1$ if $\alpha < 1$, and $\rho' = 1$ (i.e. $\rho < 1$) if $\alpha > 1$, as asserted. In the transitional case, $\alpha = 1$, equation **6.6.18** will have a double root, so that ρ and ρ' must coincide, and both be equal to unity.

As an example, consider the p.g.f. **6.6.9**, which is the p.g.f. of a *modified geometric distribution*, in which p_j falls off geometrically from $j = 1$ onwards.

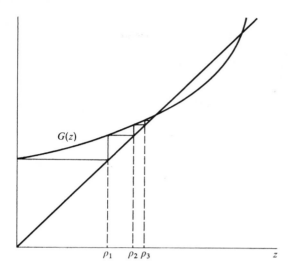

$G(z)$

p_1 p_2 p_3 z

Figure 6 A construction for the extinction probabilities of a branching process

For this case $\alpha = p + \gamma$, and **6.6.18** has the two solutions $z = 1$ and q/γ, so that

$$\rho = \min\left(1, \frac{q}{\gamma}\right).$$

Indeed, q/γ is less than unity just when $\alpha = p + \gamma$ is greater than unity.

Lotka found that the distribution of number of sons for U.S. males in 1920 was very well represented by

$p_0 = 0.4981,$

$p_j = 0.4099(0.5586)^j \quad (j = 1, 2, \ldots).$

This is just the modified geometric distribution of exercise 6.6.4, with

$$\frac{q}{1+\gamma} = 0.4981 \qquad \frac{\gamma}{1+\gamma} = 0.5586.$$

Thus

$\alpha = p + \gamma = 1.14,$

$\rho = \dfrac{q}{\gamma} = 0.89,$

so that, on our idealizing assumptions, a surname carried by just one man will ultimately disappear with probability 0.89. If it is known to be held by k men, then the probability reduces to $(0.89)^k$. So the Smiths of this world are unlikely to become extinct, despite the fact that a probability of 0.89 is relatively high.

Exercises

6.6.7 Use relation **6.6.6** to show that the sequence ρ_n is non-decreasing.

6.6.8 Complete the derivation of relations **6.6.13** and **6.6.14**.

6.6.9 Show that for the case of exercise 6.6.5

$$\alpha = +\infty$$

$$\rho = 1 - \gamma^{1/(1-\delta)}.$$

6.6.10 Suppose we modify the process by introducing immigration, so that

$$X_{n+1} = X'_{n+1} + Y_n,$$

where X'_{n+1} is the number of progeny from the X_n members of the nth generation, and Y_n is the number of immigrants. Y_n is independent of all previous
r.v.s, including X'_{n+1}, and has p.g.f. $H(z)$. Show that **6.6.6** is modified to

$$\Pi_{n+1}(z) = H(z)\Pi_n[G(z)]$$

and deduce the form of the limit p.g.f. $\Pi_\infty(z)$ (if any) if

$$G(z) = pz + q \qquad H(z) = \exp[\lambda(z-1)].$$

6.6.11 Suppose that there are two types of individuals, A and B, whose numbers in
the nth generation have joint p.g.f. $\Pi_n(z, w)$. If individuals of types A and B have
progeny p.g.f.s $G_A(z, w)$ and $G_B(z, w)$ respectively, show that

$$\Pi_{n+1}(z, w) = \Pi_n[G_A(z, w), G_B(z, w)],$$

if we make the same type of independence assumptions as in this section.

6.6.12 Suppose $X_0 = 1$, and set

$$S = \sum_0^\infty X_n.$$

This will be a finite-valued r.v. only in the case $\rho = 1$. Show that its p.g.f. $\phi(z)$
obeys the relation

$$\phi(z) = z\, G[\phi(z)]$$

and evaluate ϕ in the case **6.6.9**. This example is of interest in the theory of
polymers – organic substances whose molecules are formed by the association
of many smaller units. In some cases one can imagine that the molecule grows
by branching, and S is then the total size of the molecule. If conditions change
so that ρ becomes less than unity, then infinite values of S become possible.
This corresponds to a change of state of the polymer substance: from the 'sol'
state to the 'gel' state.

6.6.13 Let the smaller real root of **6.6.18** be denoted by ξ. Then $G'(\xi)$ will exist and $0 \leqslant G'(\xi) < 1$. Show that $\xi \geqslant p_0$. Show also that, in the range $0 \leqslant z \leqslant \xi$,

$$\xi + G'(\xi)(z - \xi) \leqslant G(z) \leqslant \xi + \frac{\xi - p_0}{\xi}(z - \xi).$$

Using these inequalities and **6.6.17**, show that

$$\left(\frac{\xi - p_0}{\xi}\right)(\xi - \rho_n) \leqslant (\xi - \rho_{n+1}) \leqslant G'(\xi)(\xi - \rho_n).$$

The first inequality indicates that, if ρ_n is not greater than ξ, then neither is ρ_{n+1}, so that ξ is the only possible limit point for the sequence $\{\rho_n\}$, and $\rho = \xi$. From the second inequality we conclude that this limit is reached exponentially fast:

$$|\rho_n - \rho| \leqslant \rho[G'(\rho)]^n.$$

7 Markov Processes

7.1 The Markov property

Consider a process for which the basic observation is of the values of a sequence of r.v.s $\{X_n\}$, so that the sample point ω corresponds to specification of the whole sequence. One often encounters such sequences when one repeatedly measures some quantity in time (such as temperature, or stock prices) and we shall for concreteness regard n as corresponding to 'time'. For some purposes we shall let n take the values $n = 0, 1, 2, \ldots$; for others it is more realistic to imagine that the process has been continuing indefinitely, so that n can take the full infinite range of values $\ldots, -2, -1, 0, 1, 2, \ldots$. In sections 7.4–7 we shall also consider processes in *continuous time*.

The simplest process would be that in which the X_n are independent r.v.s: the Bernoulli sequence has already given us an example of this. The next simplest would be that in which we suppose that specification of the value of X at any time constitutes a regeneration point of the process (see section 6.4), so that

$$E(Y_n | X_n, X_{n-1}, X_{n-2}, \ldots) = E(Y_n | X_n), \qquad 7.1.1$$

where Y_n is any r.v. defined on the future of the process at time n; i.e. a function of $X_n, X_{n+1}, X_{n+2}, \ldots$. That is, if we use the whole observed past of the process to predict the future, all the predictive value is concentrated in the last observation: if X_n is known, then the values of $X_{n-1}, X_{n-2}, \ldots$ are irrelevant for prediction. (See exercise 7.1.1.)

This is the Markov property, related to the idea of a 'complete description'. That is, the specification of the value of X_n constitutes the completest description available of the state of the process at time n, in the sense that knowledge of earlier values gives one no extra information on how the process will develop after time n.

The deterministic analogue of a Markov process would be a *first order recursion*, where X_{n+1} was determined from past values $X_n, X_{n-1}, \ldots$ by a relation

$$X_{n+1} = f(X_n, n), \qquad 7.1.2$$

depending only upon X_n and, conceivably, on n. In physical applications one frequently aims at what is by definition a complete description, by trying to express the dynamical equations as a system of first order recurrences in time (i.e. in the form **7.1.2**), or, in continuous time, as a system of first order differential equations:

$$\frac{dX(t)}{dt} = g(X(t), t). \tag{7.1.3}$$

The Markov process we are discussing is the stochastic equivalent of **7.1.2**: the continuous time Markov processes of section 7.4–7.7 are the stochastic equivalents of **7.1.3**.

A sequence $\{X_n\}$ of independent r.v.s is trivially Markov. The sequence $\{S_n; n = 1, 2, 3, \ldots\}$ of partial sums $S_n = \sum_1^n X_k$ of such a sequence is also Markov, as we leave the reader to verify. Thus, the total net winnings of a player after n independent games is a Markov variable. We shall consider other specific examples in the next section.

Exercises

7.1.1 Show that **7.1.1** has the implication

$$E(Y_n \mid X_{n_1}, X_{n_2}, X_{n_3}, \ldots) = E(Y_n \mid X_{n_1})$$

for $n \geqslant n_1 > n_2 > n_3 > \ldots$. Hence show that the *imbedded process* $\{X_{m_n}\}$ is also Markov, where $\{m_n\}$ is an increasing sequence of integers.

7.1.2 Show that **7.1.1** has the implication

$$E(Y_n Z_n \mid X_n) = E(Y_n \mid X_n) E(Z_n \mid X_n),$$

where Z_n is a r.v. defined on the past, that is a function of $X_n, X_{n-1}, X_{n-2}, \ldots$. One might express this verbally by saying that, for a Markov process, past and future are statistically independent, conditional on the present.

7.1.3 Suppose that X_n can only take a discrete set of values $\{x_k\}$ ($k = 1, 2, 3, \ldots$; all relevant n). Show that for a Markov process

$$P(X_j = x_{k_j}; j = 1, 2, \ldots, n \mid X_0 = x_{k_0}) = \prod_{j=1}^{n} P(X_j = x_{k_j} \mid X_{j-1} = x_{k_{j-1}}).$$

7.1.4 Suppose that $X_{n+1} = f(X_n, n, W_n)$, where the r.v.s W_n are mutually independent. Show that $\{X_n\}$ is Markov. In fact, any Markov process can be so represented – a nice generalization of the deterministic relation **7.1.2**.

For any process developing in time the problem of *prediction* is always a central one; for example, the prediction of X_{n+r} in terms of $X_n, X_{n-1}, X_{n-2}, \ldots,$ say. This means evaluating $E[H(X_{n+r}) \mid X_n, X_{n-1}, \ldots]$ for any function H of interest. For a Markov process, this reduces, by equation **7.1.1**, to the evaluation simply of $E[H(X_{n+r}) \mid X_n]$. It is convenient to change our time origin, and to say that we wish to determine $E[H(X_n) \mid X_m]$ for $n \geqslant m$. The following theorem gives a central recursion for this quantity, the *Chapman–Kolmogorov relation*.

Theorem 7.1.1

For a Markov process,

$$E[H(X_n)|X_m] = E[E(H(X_n)|X_r)|X_m], \qquad 7.1.4$$

for $m \leqslant r \leqslant n$. *If* X_n *can assume only a countable set of values, say* $\{x_k\}$, *then* **7.1.4** *is equivalent to*

$$P(X_n = x_k|X_m = x_j) = \sum_l P(X_r = x_l|X_m = x_j)P(X_n = x_k|X_r = x_l) \qquad 7.1.5$$

for $m \leqslant r \leqslant n$.

Equation **7.1.5** is the usual Chapman–Kolmogorov relation, and **7.1.4** is effectively a more general version of it. By breaking the average over values of $X_{m+1}, \ldots, X_n$ into an average, first with respect to $X_{r+1}, \ldots, X_n$, and then with respect to $X_{m+1}, \ldots, X_r$, we obtain

$$E[H(X_n)|X_m, X_{m-1}, \ldots] = E[E(H(X_n)|X_r, X_{r-1}, \ldots)|X_m, X_{m-1}, \ldots]. \qquad 7.1.6$$

Appealing to the Markov property **7.1.1** we see that **7.1.6** reduces to **7.1.4**.

We have not hitherto considered at all the nature of the values that X_n may assume: these can be values, not necessarily numerical, in a general space, which we shall term the *state space*. The assumption in the second part of the theorem is that the state space is *countable*. Relation **7.1.5** follows from **7.1.4** if we choose $H(X)$ as the indicator function of the set $X = x_k$, and write out all the conditional expectations explicitly.

Sometimes, instead of writing '$X_n = x_k$' we shall say 'the process is in state k at time n'; this may be more natural if X is not numerically valued. For example, if X_n were the size of a population at time n, it would be numerically valued: if it were the numbers in different age and sex categories of the population, then it would be vector-numerically valued. But the states might be simply that a man's political convictions are 'Conservative', 'Labour' or 'Liberal', or that a machine is 'functioning satisfactorily', 'due for overhaul', or 'out of action'. These states are associated with a qualitative description rather than a numerical value.

A Markov process in discrete time and with countable state space is often referred to as a *Markov chain*.

The quantities

$$p_{jk}^{mn} = P(X_n = x_k|X_m = x_j) \qquad 7.1.7$$

are known as the *transition probabilities*. The point of the Chapman–Kolmogorov relation **7.1.5** is that the transition probabilities can often be determined, by physical arguments, for $n = m+1$; relation **7.1.5** can then be used to determine p_{jk}^{mn} in general, and so solve the prediction problem. If, in particular, we take r equal to $n-1$ or $m+1$ in **7.1.5** we obtain a pair of recursions

$$p_{jk}^{mn} = \sum_l p_{jl}^{m,n-1} p_{lk}^{n-1,n}, \qquad \qquad \textbf{7.1.8}$$

$$p_{jk}^{mn} = \sum_l p_{jl}^{m,m+1} p_{lk}^{m+1,n}, \qquad \qquad \textbf{7.1.9}$$

either of which can be used to determine the p_{jk}^{mn} in terms of the one-step transition probabilities $p_{jk}^{n-1,n}$. This is entirely analogous to the determination of the sequence $\{X_n\}$ from the recursion **7.1.2** if the function $f(X, n)$ is prescribed. The two recursions **7.1.8** and **7.1.9** are known respectively as the *Kolmogorov forward* and *backward equations*; relation **7.1.2** corresponds rather to the former (see exercise 7.1.10).

One particular case of interest is the *homogeneous* case, in which $p_{jk}^{n-1,n}$ is functionally independent of n. This corresponds to the case in which the time argument n is absent from f in **7.1.2**, so that, for a given current value of X, the process always develops in the same way, regardless of the value of n. It is true in most physical situations that time has no absolute significance, and that apparent dependences on time are really reflections of the fact that one has not chosen enough variables to attain a complete description. (For example, a machine may move more easily from state 'satisfactory' to state 'faulty' as time goes on, but this is a reflection of increasing age and wear in the machine, rather than of the fact that time itself is progressing.)

Theorem 7.1.2

For the homogeneous case p_{jk}^{mn} is a function of j, k and $n-m$ alone, say $p_{jk}^{(n-m)}$. If $P^{(r)}$ is the matrix of r-step transition probabilities $(p_{jk}^{(r)})$ then the forward and backward equations can be written

$$P^{(r)} = P^{(r-1)}P \qquad P^{(r)} = PP^{(r-1)}, \qquad \qquad \textbf{7.1.10}$$

respectively, so that

$$P^{(r)} = P^r, \qquad \qquad \textbf{7.1.11}$$

where $P = P^{(1)}$, the matrix of one-step transition probabilities.

These statements are direct consequences of relations **7.1.8** and **7.1.9**, and are easily verified. For example, the intuitively obvious statement that p_{jk}^{mn} depends on m and n only via the time lag $n-m$ follows from **7.1.8** or **7.1.9** by induction on $n-m$.

Equation **7.1.11** is extremely interesting: it states that $P^{(r)}$, the r-step transition probability matrix (useful for prediction) is just the rth power of $P = P^{(1)}$. The matrix P, which we shall simply refer to as the *transition matrix*, largely characterizes the process. It will be determined by physical arguments in most applications, and so can be regarded as given.

The Markov property

Let us consider the distribution of X_n itself. By averaging relation **7.1.8** over the values of X_m we find, in the homogeneous case, that

$$P(X_n = x_k) = \sum_j P(X_{n-1} = x_j)p_{jk}. \qquad \textbf{7.1.12}$$

Now, as we shall see in the next section, for some (though far from all) Markov processes the distribution of X_n will tend to an equilibrium distribution independent of n as n increases indefinitely, so that $P(X_n = x_k) \to \pi_k$, say. If π is the column vector whose elements π_k constitute this equilibrium distribution, then we see from **7.1.12** that

$$\pi' = \pi'P, \qquad \textbf{7.1.13}$$

a system of linear equations determining π.

A Markov process for which $\lim_{n\to\infty} P(X_n = x_k | X_0 = x_j)$ exists and is independent of j is said to be *ergodic*. Equilibrium behaviour is a very important subject and we shall find examples of such processes in the next section.

Finally, we state some basic properties of P, and leave the proof largely to the reader, since the theorem can be regarded as 'optional'.

Theorem 7.1.3

A transition matrix P is characterized by the properties $p_{jk} \geqslant 0, \sum_k p_{jk} = 1$. Its eigenvalues λ lie in the unit circle, $|\lambda| \leqslant 1$. There is an eigenvalue $\lambda = 1$, with a corresponding right eigenvector $1 = (1, 1, \ldots, 1)$, and a left eigenvector π in the ergodic case.

We shall prove only the second statement. Suppose $Pz = \lambda z$, and let z_j be the element of z of maximum modulus. Then

$$|\lambda z_j| = \left| \sum_k p_{jk} z_k \right| \leqslant \sum_k p_{jk} |z_k| \leqslant |z_j|,$$

whence $|\lambda| \leqslant 1$.

Exercises

7.1.5 Show that the branching process of section 6.6 is a homogeneous Markov process with

$$G(z)^j = \sum_k p_{jk} z^k$$

and, more generally, that

$$[G^{(n)}(z)]^j = \sum_k p_{jk}^{(n)} z^k.$$

7.1.6 Show that, for an ergodic process X_m and X_n tend to independence as $|m - n| \to \infty$.

7.1.7 In view of equation **7.1.4** there is obviously an interest in functions H which are 'invariant' in the sense that

$$E[H(X_n)\mid X_{n-1}] = a_n + b_n H(X_{n-1}),$$

where a_n and b_n are constants. Show that, for the process $\{S_n\}$ mentioned on p. 129, $H(S) = S$ is such a function, and that, if $E(X_n) = 0$, then so also is $H(S) = S^2$.

For the homogeneous case, suppose that $Pz = \lambda z$, and define H by $H(x_k) = z_k$. Show that H is invariant.

7.1.8 If one pools states in a Markov process, the resulting process with a 'coarsened' state space is in general no longer Markov. However, suppose $\{A_\mu\}$ is a decomposition of the state space such that for any state j of A_μ the sum $\sum_k p_{jk}$ over all states k of A_ν is independent of j, and equal to $P_{\mu\nu}$ $(\mu, \nu = 1, 2, \ldots)$, say. Show that the process with 'states' $A_1, A_2, \ldots$ is Markov, with transition matrix $(P_{\mu\nu})$. (Cf. exercise 7.2.3.)

7.1.9 Write down P for the $\{S_n\}$ process based on Bernoulli trials.

7.1.10 Suppose that the solution of **7.1.2**, given that $X_m = \xi$, is $X_n = K(\xi; m, n)$. Determine 'forward' and 'backward' recurrences for K, analogous to the stochastic relations **7.1.8** and **7.1.9**.

7.2 Some particular Markov processes

The simplest non-trivial process is one of two states, when the transition matrix necessarily has the form

$$P = \begin{bmatrix} 1-\alpha & \alpha \\ \beta & 1-\beta \end{bmatrix}, \qquad\qquad \textbf{7.2.1}$$

α and β being the probabilities of transition. The two states may, for instance, correspond to the fact that a molecule is in one of two energy levels, or in one of two regions in space, A or B. To avoid subsequent confusion between the state of a single molecule and the state of a Markov process, we shall speak of the molecule's having two possible *positions*, A or B.

From the point of view of prediction, the quantity we need is the matrix power P^n, and we leave it to the reader to verify by induction that

$$P^n = \frac{1}{\alpha+\beta}\begin{bmatrix} \beta & \alpha \\ \beta & \alpha \end{bmatrix} + \frac{(1-\alpha-\beta)^n}{\alpha+\beta}\begin{bmatrix} \alpha & -\alpha \\ -\beta & \beta \end{bmatrix}. \qquad\qquad \textbf{7.2.2}$$

The reader who is familiar with the concept of the *spectral representation* of a matrix will recognize **7.2.2** as the spectral representation of P^n, derived from that of P.

If $|1-\alpha-\beta| < 1$, 7.2.3

then we see from **7.2.2** that

$$P^n \to \frac{1}{\alpha+\beta}\begin{bmatrix} \beta & \alpha \\ \beta & \alpha \end{bmatrix}$$ 7.2.4

as $n \to \infty$, so that the process is then *ergodic*. That is, whatever the initial distribution, the process tends to an equilibrium distribution between states, with

$$\pi_1 = \frac{\beta}{\alpha+\beta}, \qquad \pi_2 = \frac{\alpha}{\alpha+\beta}.$$

The reader can verify that this distribution satisfies the equilibrium equation **7.1.13**.

The only cases in which **7.2.2** is not fulfilled, and the process is not ergodic, are the two extreme ones $\alpha = \beta = 0$ and $\alpha = \beta = 1$. In the first case, all transition is impossible, and the molecule freezes in its initial position, whatever that was. In the second case, transition is certain, and the molecule alternates regularly between the two positions. These two cases would be referred to as 'decomposable' and 'cyclic' respectively, for obvious reasons, and represent the only two possible causes of non-ergodicity for a process with finite state space.

A modification of this process is to assume that we have N molecules, each moving independently between the two positions with transition matrix **7.2.1**. This 'expanded' process is still Markov, as we shall leave the reader to verify, with 2^N states. If we neglect the distinction between molecules, and so simply consider the number X_n of molecules at position A at time n, then $\{X_n\}$ is again a Markov process of $N+1$ states (see exercise 7.2.3). However, rather than to evaluate P and P^n for this process directly, it is simpler to note that, since the molecules move independently, the p.g.f. of X_n, conditional on the value of X_0, is given by

$$\Pi_n(z) = [(1-\alpha_n)z + \alpha_n]^{X_0}[\beta_n z + (1-\beta_n)]^{N-X_0},$$ 7.2.5

where $P^n = \begin{bmatrix} 1-\alpha_n & \alpha_n \\ \beta_n & 1-\beta_n \end{bmatrix}$ 7.2.6

is determined by **7.2.2**. In the ergodic case **7.2.4** we see that

$$\Pi_n(z) \to \left(\frac{\alpha+\beta z}{\alpha+\beta}\right)^N,$$ 7.2.7

so that the equilibrium distribution of X is binomial.

This model, known as the *Ehrenfest model*, is of considerable interest, since it was used to resolve a celebrated paradox of statistical mechanics (exercise

7.2.6). It is sometimes less reverently referred to as the 'dog-flea' model, for obvious reasons.

Most physical models are best formulated in continuous time, and these we shall consider in section 7.4. However, the genetical model of section 5.6 provides some interesting discrete time Markov processes. We assumed in section 5.6 that mating was random, but in plant and animal breeding one will have systematic methods of breeding and selection. Consider the case of *pure inbreeding*, in which an individual is mated with itself, as is possible with plants. If we follow such a breeding line, considering only a single individual in each generation, then this will constitute a Markov chain with the three states AA, Aa and aa, and transition matrix

$$P = \begin{bmatrix} 1 & \cdot & \cdot \\ \frac{1}{4} & \frac{1}{2} & \frac{1}{4} \\ \cdot & \cdot & 1 \end{bmatrix} \qquad \qquad \textbf{7.2.8}$$

We find that

$$P^n = \begin{bmatrix} 1 & & \\ \frac{1-2^{-n}}{2} & 2^{-n} & \frac{1-2^{-n}}{2} \\ & \cdot & 1 \end{bmatrix} \rightarrow \begin{bmatrix} 1 & \cdot & \cdot \\ \frac{1}{2} & \cdot & \frac{1}{2} \\ \cdot & \cdot & 1 \end{bmatrix} \qquad \textbf{7.2.9}$$

so that an AA or aa line remains unchanged, while an Aa line ultimately becomes either AA or aa, each with probability $\frac{1}{2}$. The effect of inbreeding is thus to produce a pure (homozygous) line. If the initial individual is randomly chosen from a population in which the three types occur in proportions p_0, q_0, r_0, then the probability that the line ultimately tends to the AA strain is $p_0 + \frac{1}{2}q_0 = \theta$, which is the A-gene frequency in the initial population.

Suppose we consider *inbreeding with selection against a*, so that all aa individuals are discarded, and one continues with breeding in a given generation until a non-aa individual has been produced. The process is still Markov, with only two states, AA and Aa, and transition matrix

$$P = \begin{bmatrix} 1 & \cdot \\ \frac{1}{3} & \frac{2}{3} \end{bmatrix},$$

for which

$$P^n = \begin{bmatrix} 1 & \cdot \\ 1-(\frac{2}{3})^n & (\frac{2}{3})^n \end{bmatrix} \rightarrow \begin{bmatrix} 1 & \cdot \\ 1 & \cdot \end{bmatrix}$$

One thus ends with a pure AA line under all circumstances.

If p_n, q_n, r_n are the proportions of the three types in the nth generation, we found that these obeyed the *nonlinear* recursion **5.6.14–16** for the case of random mating. This is because, in this case, the three genotypes cannot be considered as Markov states; the mating process is such that one has to refer back to the constitution of the population.

For an adequate treatment of this model as a Markov process, one would have to prescribe a state by the actual *numbers* of the three genotypes in the population at the generation.

Exercises

7.2.1   Suppose that $\{X_n\}$ is a numerically valued Markov process with transition matrix **7.2.1**. Show that, for the ergodic case, the correlation coefficient between X_n and X_{n+s} is $(1 - \alpha - \beta)^s$, if s is positive, and n so large that equilibrium has been attained.

7.2.2   What are the transition probabilities for the Ehrenfest model (a) where a distinction is made between molecules, and (b) where it is not?

7.2.3   Show (cf. exercise 7.1.7) that the Markov character of the Ehrenfest model is not lost when the distinction between the molecules is abandoned, the 2^N states being grouped into $N + 1$ states.

7.2.4   For the Ehrenfest model, let $\Pi(z, w)$ be the joint p.g.f. of the numbers of molecules in the two positions. Show that

$$\Pi_{n+1}(z, w) = \Pi_n[(1 - \alpha)z + \alpha w, \beta z + (1 - \beta)w].$$

Note that this is a special case of exercise 6.6.11.

7.2.5   Use relation **6.3.21** to show that the expected recurrence time to state $X = j$ of the Ehrenfest model is $\left[\binom{N}{j} \frac{\alpha^j \beta^{N-j}}{(\alpha + \beta)^N} \right]^{-1}$. Consider how this varies with j in the case $\alpha = \beta$.

7.2.6   Consider the Ehrenfest model in the ergodic case **7.2.3**. Show that X_n and X_{n+s} have an equilibrium distribution with joint p.g.f.

$$\Pi(z, w) = \lim_{n \to \infty} E(z^{X_n} w^{X_{n+s}})$$

$$= \left[\frac{(\alpha + \beta z)(\alpha + \beta w) + \alpha\beta(1 - \alpha - \beta)^s(1 - z)(1 - w)}{(\alpha + \beta)^2} \right]^N.$$

From the symmetry of this expression it follows that in equilibrium

$$P(X_n = j, X_{n+s} = k) = P(X_n = k, X_{n+s} = j).$$

The point about this result is that it demonstrates a symmetry in the process between past and future, whereas the generally valid inequality

$$P(X_{n+s} = k \mid X_n = j) \neq P(X_{n+s} = j \mid X_n = k)$$

corresponds to a tendency for the process to move towards the more probable states as time goes on. These two viewpoints of the same process reconcile the time-reversibility of classical mechanics, and the irreversibility of thermo-dynamics.

7.2.7 Consider a brother–sister mating system, in which, conditional on the geno-types of the parents, brother and sister receive their genes independently according to the rules of section 5.6. Take the genotypes of brother and of sister as description of the state (giving nine states), and write down the transition matrix for the process. Prove that this system of breeding always produces a homozygous line ultimately.

7.3 The simple random walk

Consider a Markov process with states $j = \ldots, -1, 0, 1, \ldots$ for which the transition probabilities are given by

$$p_{jk} = \begin{cases} p_j & (k = j+1), \\ r_j & (k = j), \\ q_j & (k = j-1), \\ 0 & (\text{otherwise}). \end{cases} \qquad \textbf{7.3.1}$$

Since $\sum_k p_{jk} = 1$, we must have $p_j + q_j + r_j = 1$. If òne visualizes j as the co-ordinate of a point on a line, then it is as though the point representing the state of the process can move either one step to the right with probability p_j, one step to the left with probability q_j, or stay where it is with probability r_j; hence the term *random walk*.

The model has actually been used to represent Brownian motion, that is the motion of a large colloidal particle subject to random forces. It is perhaps ‚not really adequate for this (see section 12.4) but it serves as a first approxima-tion.

However, its classical origin is in the theory of games of chance. Suppose two players A and B play a series of games of chance, and that at each game the winner takes unit stake from the loser. Then p_j, q_j and r_j could be taken as the probabilities that A will respectively win, lose or draw, if he holds a capital of j units. In general these probabilities will be largely independent of j, but it is useful to retain the formal dependence. For suppose the rule of the game is that play must cease if $j = 0$ because then A is ruined. Then one remains in the state $j = 0$ (which is then an *absorbing state*) and it is as though $p_0 = q_0 = 0$ and $r_0 = 1$. Similarly, if a is the total capital of the two players, then play must equally stop at $j = a$, when B is ruined, so that $p_a = q_a = 0$ and $r_a = 1$.

The systems of forward and backward equations are respectively

$$p_{jk}^{(n)} = p_{j,k-1}^{(n-1)} p_{k-1} + p_{jk}^{(n-1)} r_k + p_{j,k+1}^{(n-1)} q_{k+1}, \qquad \textbf{7.3.2}$$

$$p_{jk}^{(n)} = p_j p_{j+1,k}^{(n-1)} + r_j p_{jk}^{(n-1)} + q_j p_{j-1,k}^{(n-1)}. \qquad \textbf{7.3.3}$$

The system **7.3.2** is the more useful if we are considering $p_{jk}^{(n)}$ for a fixed initial state j and a variable current state k; the system **7.3.3** is the more useful in the converse situation.

An important example of the converse situation occurs in the gaming problem, if we wish to calculate

$$A_j = \lim_{n \to \infty} p_{j,0}^{(n)}, \tag{7.3.4}$$

the *probability that player A is ultimately ruined*, given that he starts with an initial capital j. If we assume that the win probabilities are independent of j in the range $j = 1, 2, \ldots, a-1$, then we find from **7.3.3** that

$$A_j = pA_{j+1} + rA_j + qA_{j-1} \quad (j = 1, 2, \ldots, a-1), \tag{7.3.5}$$

$$A_0 = 1, \tag{7.3.6}$$
$$A_a = 0.$$

The difference equation **7.3.5** has solutions of the form $A_j = z^j$: inserting this trial solution in **7.3.5** we find that the possible values are $z = 1$ and q/p. Choosing the linear combination of these two solutions which satisfies the boundary conditions **7.3.6**, we find that

$$A_j = \frac{(q/p)^j - (q/p)^a}{1 - (q/p)^a}. \tag{7.3.7}$$

One special case of interest is that in which $a \to \infty$, so that A is playing against an infinitely rich opponent. We find then from **7.3.7** that

$$A_j = \begin{cases} (q/p)^j & (p > q), \\ 1 & (p \leq q). \end{cases} \tag{7.3.8}$$

Thus A loses with probability one even in the case $p = q$, when the game is a fair one. If the game is advantageous to him ($p > q$) his chance of ruin decreases exponentially with the size of his original capital. These results are reminiscent of those for the branching process (section 6.6), and indeed the two can be related.

A rather more developed model of this type is used to calculate ruin probabilities for insurance companies, 'wins' then being the inflow of premiums, 'losses' the outflow of claims, and j corresponding to the initial risk capital available. The company is effectively playing against an infinitely rich opponent, since it never enters an absorbing state of 'victory', no matter how large its capital.

Exercises

7.3.1 Use equations **7.3.3** to show that, if M_j is the expected duration of the game conditional on a start from j, then

$$M_j = 1 + pM_{j+1} + rM_j + qM_{j-1} \quad (j = 1, 2, \ldots, a-1)$$

with $M_0 = M_a = 0$, and solve for M_j.

7.3.2   The solution of **7.3.5** with boundary condition $A_0 = A_a = 1$ is $A_j = 1$, which proves that termination (i.e. absorption at either 0 or a) is certain, at least if a is finite. For the case $p \neq q$ we can find a useful lower bound to the rate of absorption. Show by induction from **7.3.2** that

$$p_{jk}^{(n)} \leqslant z^{k-j} \rho^n \quad (k = 1, 2, \ldots, a-1),$$

where z is the positive value minimizing

$$\rho = pz + r + qz^{-1}.$$

7.3.3   Note that in exercise 6.4.7 we effectively obtained, not only the result **7.3.8**, but also the p.g.f. of time to absorption. Generalize the arguments of the present section to obtain this p.g.f.

7.3.4   Solve for the A_j of the text in the case where p, q and r are j-dependent; the states $j = 0$, a still being absorbing.

7.3.5   Suppose we use model **7.3.1** to represent the Brownian motion of a particle in a one-dimensional container with 'walls' at $j = 0$ and a, so that $q_0 = p_a = 0$. Show that, if π_j is the equilibrium probability that the particle is at position j, then

$$\pi_j = \frac{p_0 p_1 \cdots p_{j-1}}{q_1 q_2 \cdots q_j} \pi_0 \quad (j = 1, 2, \ldots, a).$$

7.4   Markov processes in continuous time

Most physical processes are more naturally formulated in continuous than in discrete time, so that it is natural to look for a continuous time version of a Markov process. We shall continue to assume for convenience that the variable X_t is discrete (i.e. that state space is countable), taking values x_k. We shall also assume the process time-homogeneous, so that we can write

$$P(X_{s+t} = x_k \mid X_s = x_j) = p_{jk}(t) \qquad \text{7.4.1}$$

for $t \geqslant 0$.

The Markov property, as expressed in

$$E(Y_t \mid X_s; s \leqslant t) = E(Y_t \mid X_t), \qquad \text{7.4.2}$$

where Y_t is any r.v. defined in terms of X_s ($s \geqslant t$), leads, as in Theorem 7.1.1, to the Chapman–Kolmogorov equation

$$p_{jk}(s+t) = \sum_m p_{jm}(s) p_{mk}(t) \quad (s, t \geqslant 0). \qquad \text{7.4.3}$$

We can rewrite this equation in matrix form

$$P(s+t) = P(s)P(t),$$ **7.4.4**

where $P(t)$ is the matrix of t-step transition probabilities.

Previously, we regarded the one-step transition probabilities as given, and used **7.4.4** to derive the n-step probabilities from them. Presumably the analogue must now be to let s or t tend to zero in **7.4.4**, and thus obtain continuous time analogues of the backward and forward equations.

We must have $p_{jk}(0) = \delta_{jk}$, i.e. $P(0) = I$, as a matter of definition. We can probably expect that in most cases the stronger relation

$$P(t) \to I \quad \text{as} \quad t \downarrow 0$$

will hold, because one will expect that the probability that the process has left its initial state x_k in time t will tend to zero as t does.

In fact, we shall make an even stronger assumption: that $P(t)$ possesses a right-derivative Q at $t = 0$, so that

$$p_{jk}(t) = \delta_{jk} + q_{jk}t + o(t) \quad (t \geq 0).$$ **7.4.5**

It is not necessary that a continuous time Markov process has this property, but almost all processes of physical interest do. If we consider a r.v. W_t which takes value w_k with probability $p_{jk}(t)$ at time t, one would expect in most cases that $E(W_t) = \sum_k p_{jk}(t)w_k$ would be time-differentiable, and the corresponding demand for the probabilities follows as a special case.

Since $p_{jk}(0) = 0$ if $j \neq k$, and $p_{jk}(t) \geq 0$, we must then have $q_{jk} \geq 0$. If $j \neq k$ then q_{jk} is known as the *probability intensity of transition from state j to state k*, and can be regarded as the rate in time with which such transitions occur, given that one is in state j. By the same argument, $q_{jj} \leq 0$. In fact, since $\sum_k p_{jk} = 1$, then $\sum_k q_{jk} = 0$. It is sometimes useful to introduce the quantity

$$q_j = -q_{jj} = \sum_{k \neq j} q_{jk},$$ **7.4.6**

which is the total intensity of transition out of state j; i.e. the rate at which state j is left for some other state.

By letting t and s tend to zero respectively in **7.4.4** we obtain the forward and backward equations

$$\dot{P}(t) = P(t)Q,$$ **7.4.7**

$$\dot{P}(t) = Q\,P(t),$$ **7.4.8**

where $\dot{P}(t) = dP(t)/dt$. Written in full, these read

$$\dot{p}_{jk}(t) = \sum_{m \neq k} p_{jm}(t)q_{mk} - q_k p_{jk}(t),$$ **7.4.9**

$$\dot{p}_{jk}(t) = \sum_{m \neq j} q_{jm} p_{mk}(t) - q_j p_{jk}(t).$$ **7.4.10**

Note an implication of equation **7.4.10**; that the derivatives $\dot{p}_{jk}(t)$ exist for positive t if they exist for $t = 0+$. We single out **7.4.10** rather than **7.4.9**, because the sum in the backward equation is always convergent (since it is less than $\sum_m q_{jm} = q_j$), while that in the forward equation may not be.

Exercises

7.4.1   Show that the eigenvalues of Q all have non-positive real parts, and that zero is an eigenvalue. Also that 1 is a right eigenvector for the zero eigenvalue, and the equilibrium distribution vector π (if there is one) is a left eigenvector.

7.4.2   Confirm that equations **7.4.7** and **7.4.8** have a formal solution

$$P(t) = \sum_0^\infty \frac{(tQ)^r}{r!} = e^{tQ}.$$

7.4.3   Determine $P(t)$ explicitly in the case

$$Q = \begin{bmatrix} -\lambda & \lambda \\ \mu & -\mu \end{bmatrix}$$

(cf. the analagous discrete time case **7.2.1** and **7.2.2**).

7.5   The Poisson process in time

Consider the case when one moves up through the states at a constant rate, so that

$$q_{jk} = \begin{cases} \lambda & (k = j+1), \\ -\lambda & (k = j), \\ 0 & (\text{otherwise}). \end{cases} \qquad \textbf{7.5.1}$$

This is the case of the Poisson process in time, where one considers a variable $n(t)$, the number of 'events' which have taken place by time t, and we make the identification

$$p_{jk}(t) = P[n(t) = k \,|\, n(0) = j]. \qquad \textbf{7.5.2}$$

For example, $n(t)$ may be the number of registrations on a Geiger counter, the number of children born, or the number of calls entering a telephone exchange. Assumptions tacit in our model are: (i) that occurrence of events after time t is not conditioned by events which have occurred up to time t (so, for example, we must assume that one telephone call cannot block another); (ii) that events occur with constant intensity λ; and (iii) that multiple events (e.g. double registrations on a counter, or the arrival of twins) occur with zero intensity (but see exercise 7.5.1). These are in fact equivalent to our earlier specification of a Poisson process (section 4.7), to which we shall return later.

If we assume that $n(t)$ is the number of events in $(0, t)$, so that $n(0) = 0$, and set $p_{0k}(t) = P_k(t)$, then the forward equations **7.4.9** become

$$\dot{P}_0 = -\lambda P_0,$$
$$\dot{P}_k = \lambda P_{k-1} - \lambda P_k \quad (k = 1, 2, \ldots), \qquad \textbf{7.5.3}$$

with the initial condition $P_k(0) = \delta_k$. These can be solved recursively, and one finds that

$$P_k = \frac{e^{-\lambda t}(\lambda t)^k}{k!}, \qquad \textbf{7.5.4}$$

which is the Poisson distribution one might have expected.

Actually, a more economical way of solving the infinite equation system **7.5.3** is to work in terms of the p.g.f.

$$\Pi(z, t) = E(z^{n(t)}) = \sum_k P_k(t) z^k. \qquad \textbf{7.5.5}$$

Multiplying the kth equation of **7.5.5** by z^k, and adding, we find that

$$\frac{\partial \Pi(z, t)}{\partial t} = \lambda(z - 1)\Pi(z, t), \qquad \textbf{7.5.6}$$

with the initial condition $\Pi(z, 0) = 1$. Equation **7.5.6** can be integrated immediately to give

$$\Pi(z, t) = e^{\lambda t(z - 1)} \qquad \textbf{7.5.7}$$

consistently with **7.5.4**. This is a technique we shall employ many times, because the expectation defined by **7.5.5** is one that develops naturally in time for many of the processes with which we shall be dealing.

Our original formulation of the Poisson process (section 4.7) is perhaps a more powerful one, because it enables us to see the process as possibly multi-dimensional (the differential formulation **7.5.3** is essentially a one-dimensional one), and to arrive quickly at more general expectations. This is because it takes the rather more natural view of looking at the distribution of particles over space (or events over time) rather than at the number of particles in a given space element.

Suppose, then, we begin by considering the case of N particles uniformly and independently distributed on an interval $(0, T)$, so that

$$\Pi(z) = E\left[\prod_1^n z(s_j) \right] = \left[T^{-1} \int_0^T z(s)\, ds \right]^N, \qquad \textbf{7.5.8}$$

where s_j is the coordinate of the jth particle, and $z(s)$ is an arbitrary function.

Note that we can also write the expectation **7.5.8** as

$$E\left[\exp\left(\sum_1^N \log z(s_j)\right)\right] = E\left[\exp\left(\int_0^T \log z(s)\, dn(s)\right)\right] \tag{7.5.9}$$

so that $\Pi(z)$ is something rather like a joint p.g.f. of all the increments $dn(s)$.

Suppose now that $z(s)$ tends to unity with increasing s sufficiently fast that

$$\int_0^\infty |z(s)-1|\, ds < \infty. \tag{7.5.10}$$

Letting N and T tend to infinity in **7.5.8** in such a way that $N/T \to \lambda$, we find that

$$\Pi(z) \to \lim_{T\to\infty}\left[1 + T^{-1}\int_0^T (z(s)-1)\, ds\right]^{\lambda T}$$

$$= \exp\left[\lambda \int_0^\infty (z(s)-1)\, ds\right]. \tag{7.5.11}$$

This is what we mean by saying that we can quickly obtain more general expectations from the original formulation. The same calculation can obviously be carried out for the whole axis rather than the half-axis, or for an infinite region in several dimensions.

Choosing $z(s) = z$ $(0 \leqslant s \leqslant t)$ and $z(s) = 1$ (elsewhere), we find that **7.5.11** reduces to **7.5.7**.

Exercises

7.5.1   Suppose the Poisson process is modified to allow multiple events, so that

$$q_{jk} = \begin{cases} \lambda_{k-j} & (k > j), \\ 0 & (k < j), \end{cases}$$

where λ_r is the intensity of an r-fold event (e.g. $\lambda_2, \lambda_3, \ldots$ are the respective birth intensities of twins, triplets, etc.). Show that the p.g.f. of the number of events in $(0, t)$ is

$$\Pi(z, t) = \exp\left[t \sum_r \lambda_r(z^r - 1)\right]$$

7.5.2   Suppose particles carrying unit positive and negative charge enter a container at respective rates (probability intensities) λ and μ. Show that the p.g.f. of the net positive charge accumulated after time t is

$$\exp[t(\lambda z + \mu z^{-1} - \lambda - \mu)].$$

Derive a series expression for $P_k(t)$.

7.6 Birth processes

The general process in which the only possible transitions are of the form $j \to j+1$:

$$q_{jk} = \begin{cases} \lambda_j & (k = j+1), \\ -\lambda_j & (k = j), \\ 0 & (\text{otherwise}) \end{cases} \qquad \textbf{7.6.1}$$

is termed a *birth process*. The forward equations are

$$\dot{p}_{jk}(t) = \lambda_{k-1} p_{j,k-1}(t) - \lambda_k p_{jk}(t). \qquad \textbf{7.6.2}$$

The Poisson process is a particular case; another one is that of radioactive transmutation, where a radioactive atom can pass through a number of states by successive disintegrations. We assume that the process is one of degradation, so that return to an earlier state is never possible.

Suppose we use $k = 0$ to label the first state of the series, and that $k = m$ is the final one. If the atom began in state 0 at $t = 0$, and $P_k(t)$ is the probability that it is in state k at time t, then

$$\left. \begin{aligned} \dot{P}_0 &= -\lambda_0 P_0, \\ \dot{P}_k &= \lambda_{k-1} P_{k-1} - \lambda_k P_k \quad (k = 1, 2, \ldots, m-1), \\ \dot{P}_m &= \lambda_{m-1} P_{m-1}, \end{aligned} \right\} \qquad \textbf{7.6.3}$$

with $P_k(0) = \delta_k$.

This system can be solved recursively whatever the λ_k; we find for example that

$$P_0(t) = e^{-\lambda_0 t},$$

$$P_1(t) = \frac{\lambda_0}{\lambda_1 - \lambda_0}(e^{-\lambda_0 t} - e^{-\lambda_1 t}). \qquad \textbf{7.6.4}$$

A general technique of evaluation is indicated in exercise 7.6.1. Note that in solving **7.6.3** we can also solve the case where n atoms decay independently from the initial state $j = 0$ at $t = 0$. Because, in virtue of independence, the multinomial distribution will hold, and

$$P(N_k(t) = n_k; k = 1, 2, \ldots, m) = n! \prod_k \frac{P_k(t)^{n_k}}{n_k!} \qquad \textbf{7.6.5}$$

where $N_k(t)$ is the number of atoms in state k at time t.

An important special case of **7.6.1** is the *simple birth process*, when λ_j is proportional to j, say $\lambda_j = \lambda j$. This comes about as follows. Consider a colony of j amoebae, each of which can reproduce by simple division with intensity λ, and suppose that individuals reproduce independently. Then the probability of one division in time t is, by the binomial theorem,

$$\binom{j}{1}[\lambda t + o(t)][1 - \lambda t + o(t)]^{j-1} = \lambda j t + o(t),$$

so that the transition $j \to j+1$ does indeed have intensity λj. The probability of more than one division must be of order $(\lambda t)^2$ at least, so that transitions $j \to j+r$ have zero intensity if $r > 1$.

The system **7.6.2** for this case can be solved in a number of ways. For example, it can be solved recursively, as in **7.6.4**, and with a little ingenuity one can see the general solution. Alternatively, one can introduce the p.g.f.

$$\Pi_j(z, t) = E(z^{n(t)} | n(0) = j)$$

$$= \sum_k p_{jk}(t) z^k, \qquad\qquad \textbf{7.6.6}$$

and obtain from **7.6.2** the equation

$$\frac{\partial \Pi_j}{\partial t} = \lambda z(z-1) \frac{\partial \Pi_j}{\partial z}. \qquad\qquad \textbf{7.6.7}$$

This is a linear partial differential equation, with initial condition $\Pi_j(z, 0) = z^j$, which can be solved by standard techniques. However, we can avoid even this if we note that

$$\Pi_j(z, t) = [\Pi_1(z, t)]^j \qquad\qquad \textbf{7.6.8}$$

because Π_j is the p.g.f. of the sum of j independent and identically distributed variables, the progeny from each of the j initial members of the population. The backward equation **7.4.10** gives us, in terms of the p.g.f.s

$$\frac{\partial \Pi_j}{\partial t} = \lambda_j(\Pi_{j+1} - \Pi_j), \qquad\qquad \textbf{7.6.9}$$

which, in virtue of the relation **7.6.8**, yields

$$\frac{\partial \Pi_1}{\partial t} = \lambda(\Pi_1^2 - \Pi_1), \qquad\qquad \textbf{7.6.10}$$

with initial condition $\Pi_1(z, 0) = z$. This is readily solved, and we find that

$$\Pi_j(z, t) = \left[\frac{ze^{-\lambda t}}{1 - z(1 - e^{-\lambda t})} \right]^j, \qquad\qquad \textbf{7.6.11}$$

which is a negative binomial distribution.

We see from **7.6.11** that

$$E[n(t)] = je^{\lambda t}, \qquad\qquad \textbf{7.6.12}$$

$$\text{var}[n(t)] = je^{\lambda t}(e^{\lambda t} - 1). \qquad\qquad \textbf{7.6.13}$$

Relation **7.6.12** is just what one might expect, since our process is a stochastic version of the deterministic model $\dot{n} = \lambda n$. However, we see from **7.6.13** that the random variations (as measured by the standard deviation) are of the same order as the mean value, so the stochastic element affects behaviour profoundly.

The simple birth process has been used not only as a model for simple biological populations, but also in the study of cosmic rays, to describe the number of particles in a 'shower' energized by a single parent energetic particle.

Exercises

7.6.1　Define

$$\bar{P}_k(u) = \int_0^\infty P_k(t)e^{-ut}\, dt,$$

where the $P_k(t)$ are the probabilities of **7.6.3**. Show that

$$\bar{P}_k(u) = \frac{\lambda_0 \lambda_1 \dots \lambda_{k-1}}{(\lambda_0+u)(\lambda_1+u)\dots(\lambda_k+u)}.$$

7.6.2　Show that the equation system **7.6.9** *implies* conclusion **7.6.8**.

7.6.3　One can also define a *simple death process*, for which

$$q_{jk} = \begin{cases} \mu j & (k = j-1), \\ -\mu j & (k = j), \\ 0 & (\text{otherwise}). \end{cases}$$

Show that for this

$$\Pi_j(z, t) = [1 + e^{-\lambda t}(z - 1)]^j$$

and interpret.

7.6.4　Show that the times spent in states $j, j+1, \dots$ of a birth process are independent exponential variables with means $\lambda_j^{-1}, \lambda_{j+1}^{-1}, \dots$.

7.6.5　Show that the mean time needed for the population in a simple birth process to increase from j to k is asymptotic to $\lambda^{-1} \log(k/j)$ for large k.

7.6.6　*A model of molecular dissociation (Whittle).* Consider a situation in which j molecules can be linked in an open chain to form what we shall call a *j-mer*. Suppose that any link of any *j*-mer can break independently, with intensity λ. Consider

$$\Pi_j(z, t) = E\left(\prod_1^\infty z_k^{n_k(t)} \mid n_k(0) = \delta_{jk}\right),$$

the joint p.g.f. of the number of k-mers at time t, $n_k(t)$ $(k = 1, 2, \dots)$, given that we started with just one j-mer. Show that

$$\frac{\partial \Pi_j}{\partial t} = \lambda \sum_{k=1}^{j-1} \Pi_k \Pi_{j-k} - \lambda(j-1)\Pi_j$$

with $\Pi_j(z, 0) = z_j$, and hence that Π_j is the coefficient of ξ^j in the expansion of

$$\frac{\sum_1^\infty z_k \xi^k e^{-\lambda(k-1)t}}{1 - (1 - e^{-\lambda t}) \sum_1^\infty z_k \xi^k e^{-\lambda(k-1)t}}$$

in powers of ξ. Interpret the result in the cases $z_j = z$, $z_j = z^j$.

7.7 Birth and death processes

A birth and death process is one in which the only possible transitions are $j \to j \pm 1$:

$$q_{jk} = \begin{cases} \lambda_j & (k = j+1), \\ -(\lambda_j + \mu_j) & (k = j), \\ \mu_j & (k = j-1), \\ 0 & (\text{otherwise}). \end{cases} \qquad \textbf{7.7.1}$$

Thus it is just the continuous time analogue of the simple random walk discussed in section 7.3.

The forward equations are

$$\dot{p}_{jk} = \lambda_{k-1} p_{j,k-1} + \mu_{k+1} p_{j,k+1} - (\lambda_k + \mu_k) p_{jk}. \qquad \textbf{7.7.2}$$

The equilibrium behaviour of such processes can be interesting – in the case of a pure birth process one could only either progress to infinity (as in the Poisson process) or end in an absorbing state (as in radioactive transmutation). The equations for the equilibrium probabilities π_k (supposing these to exist) are

$$\lambda_{k-1} \pi_{k-1} + \mu_{k+1} \pi_{k+1} = (\lambda_k + \mu_k)\pi_k. \qquad \textbf{7.7.3}$$

Suppose that $\mu_0 = 0$, so that passage into states $k < 0$ is impossible. We find then, by considering 7.7.3 for $k = 0, 1, \ldots$, that the simpler equation

$$\lambda_{k-1} \pi_{k-1} = \mu_k \pi_k \quad (k = 1, 2, \ldots) \qquad \textbf{7.7.4}$$

holds, whence

$$\pi_k = \frac{\lambda_0 \lambda_1 \ldots \lambda_{k-1}}{\mu_1 \mu_2 \ldots \mu_k} \pi_0 \quad (k = 1, 2, \ldots). \qquad \textbf{7.7.5}$$

Relation 7.7.4 makes sense upon examination: it states that the 'probability flux' from $k-1$ to k must balance that from k to $k-1$. Such *detailed balance* relations are invaluable, when they can be found.

As an example, let k denote the number of customers in a queue, and suppose that customers leave the queue with intensity μ (for $k > 0$, of course) and join it with intensity k ($k < a$) or zero ($k \geqslant a$). We are thus supposing that, for some reason, the queue size must not exceed a, and potential customers simply go elsewhere if they find it this large. By applying **7.7.5** and using the normalization $\sum_{0}^{a} \pi_k = 1$, we see that

$$\pi_k = \frac{(1-\rho)\rho^k}{1-\rho^{a+1}} \quad (k = 0, 1, \ldots, a), \qquad \textbf{7.7.6}$$

where $\rho = \lambda/\mu$, the so-called *traffic intensity*. If we now remove the limitation on queue size, and let $a \to \infty$, we see that there are two possible types of behaviour. If $\rho < 1$, expression **7.7.6** converges to a perfectly acceptable geometric distribution. However, if $\rho \geqslant 1$, the distribution becomes concentrated onto indefinitely large values of k. This is the case when customers are joining the queue faster than they can be served, on average, so that the queue size builds up indefinitely.

One model of special interest is the *simple birth and death process*, for which $\lambda_j = \lambda j$, $\mu_j = \mu j$. As for the simple birth process, one finds this form for the transition intensities if one considers for example, a biological population in which each member can reproduce or die with individual intensity λ or μ respectively, and where these events are independent for distinct individuals. Adopting the same definition **7.6.6** as before, we see that **7.6.8** still holds, and that equation **7.6.10** is now modified to

$$\frac{\partial \Pi_1}{\partial t} = (\lambda \Pi_1 - \mu)(\Pi_1 - 1). \qquad \textbf{7.7.7}$$

Solving equation **7.7.7** with the initial condition $\Pi_1(z, 0) = z$, we find that

$$\Pi_j(z, t) = \left[\frac{\mu(1-z) - (\mu - \lambda z)e^{-(\lambda-\mu)t}}{\lambda(1-z) - (\mu - \lambda z)e^{-(\lambda-\mu)t}} \right]^j. \qquad \textbf{7.7.8}$$

The model is a rather oversimplified one for biological populations, because it neglects effects such as age, sex and limiting environmental factors. However, it is useful as a first approximation for populations of simple organisms. It has also been used as a model for energetic particle showers.

The probability that the population is extinct by time t is

$$P_0(t) = \Pi_j(0, t) = \left[\frac{\mu - \mu e^{-(\lambda-\mu)t}}{\lambda - \mu e^{-(\lambda-\mu)t}} \right]^j. \qquad \textbf{7.7.9}$$

For large t this tends to 1 if $\lambda \leqslant \mu$, and to $(\mu/\lambda)^j$ if $\lambda > \mu$, so the conclusions are much as for the branching process of section 6.6: ultimate extinction is certain if and only if the net birth rate is less than or equal to zero. Indeed, the simple birth and death process is a continuous time version of a branching process, with a progeny distribution over a positive time which, as we see from **7.7.8**, is modified geometric.

Exercises

7.7.1   The birth–death process is a stochastic analogue of a deterministic process $\dot{n} = \lambda_n - \mu_n$. The deterministic process would have possible equilibria at n-values determined by $\lambda_n = \mu_n$, these being locally stable or unstable according as $\lambda_n - \mu_n$ is decreasing or increasing in n. Show from equation **7.7.5** that P_n correspondingly has a local maximum or minimum, respectively.

7.7.2   A telephone exchange has a channels; suppose that j of these are engaged. Suppose that $j \to j-1$ with intensity μj, and that $j \to j+1$ with intensity λ if $j < a$, or zero if $j \geqslant a$, when the exchange is saturated and any incoming call is lost. Calculate the probability that the exchange is saturated.

7.7.3   Consider the birth–death–immigration process with $\lambda_j = \lambda j + v$, $\mu_j = \mu j$. Show from the backward equations that $\Pi_j(z, t)$ is of the form $\phi(z, t)^j \psi(z, t)$; interpret the expression and determine ϕ and ψ. Under what conditions will an equilibrium distribution exist, and what is it? Specialize to the case $\lambda = 0$.

8 Continuous Distributions

8.1. Distributions with a density

We have hitherto largely considered processes in which the sample space is *discrete*: that is, in which an experiment has at most countably many outcomes. However, it is plain that we shall often wish to deal with r.v.s which may vary continuously (e.g. position, velocity, energy, rainfall), and so we must work with more general sample spaces.

For the discrete case the expectation functional is given by the sum **2.4.1**: for the general case it is given by some very general type of integral **3.2.5** (although it is central to our approach that the known expectations will not necessarily be given in this form).

However, in this chapter we shall largely restrict ourselves to the so-called *absolutely continuous* case, when a *probability density* exists, and integral **3.2.5** is of a relatively elementary type.

Suppose that an experiment consists in the observation of the values of n r.v.s $X_1, X_2, \ldots, X_n$, so that Ω can be taken as R^n, an n-dimensional Euclidean space. Suppose also that we know the expectation of a function to be given by the integral

$$E[H(X_1, X_2, \ldots, X_n)]$$

$$= \int\int \ldots \int H(x_1, x_2, \ldots, x_n) f(x_1, x_2, \ldots, x_n) \, dx_1 \, dx_2 \ldots dx_n \qquad \textbf{8.1.1}$$

for a class of functions H which includes at least the *simple functions* (i.e. functions for which Ω can be divided into a finite number of n-dimensional intervals

$$a_j < X_j \leqslant b_j \quad (j = 1, 2, \ldots, n),$$

such that H is constant on each of these intervals). Then f will be known as the *probability density* of $X_1, X_2, \ldots, X_n$ (cf. section 2.4).

Very often we shall write integral **8.1.1** in the vector form

$$E[H(X)] = \int H(x) f(x) \, dx, \qquad \textbf{8.1.2}$$

where X is the vector with elements $X_1, X_2, \ldots, X_n$, and $dx = dx_1 \, dx_2 \ldots dx_n$.

Theorem 8.1.1

If A is a set in Ω consisting of a finite union of intervals, then

$$P(A) = P(X \in A) = \int_A f(x)\,dx. \qquad \textbf{8.1.3}$$

This has as consequence

$$f(x) = \frac{\partial^n F(x)}{\partial x_1 \partial x_2 \dots \partial x_n}, \qquad \textbf{8.1.4}$$

where $F(x)$ is the distribution function *of X*:

$$F(x) = P(X \leqslant x). \qquad \textbf{8.1.5}$$

Since the indicator function of A is certainly a simple function, equation **8.1.3** follows from **8.1.2**. The formula **8.1.4** (which shows that f is determined by **8.1.2** if the values $E(H)$ are actually known for the class of functions indicated) follows by differentiation of the special case of **8.1.2**:

$$F(x) = \int_{-\infty}^{x_1} dy_1 \dots \int_{-\infty}^{x_n} dy_n\, f(y_1, \dots, y_n). \qquad \textbf{8.1.6}$$

We may recall other consequences of equation **8.1.3**: that f is non-negative almost everywhere, and integrates to unity (section 2.4).

By appealing to the axioms, we shall see in section 11.1 that the validity of **8.1.2** can be extended to the case when H is a *Borel function* (a limit of simple functions) and that **8.1.3** can be similarly extended to the case where A is the set of x satisfying $H(x) \geqslant 0$, for some Borel function H. This class of functions and sets (which are respectively included in the class of *integrable* functions and *measurable* sets) is wide enough for all practical purposes, and we shall assume henceforth (tacitly, unless emphasis is needed) that all r.v.s and events considered fall in these classes.

Theorem 8.1.2

*If **8.1.1** holds then the smaller set of r.v.s $X_1, X_2, \dots, X_m$ ($m \leqslant n$) also possesses a probability density, determined by*

$$f(x_1, x_2, \dots, x_m) = \int\int \dots \int f(x_1, x_2, \dots, x_n)\, dx_{m+1}\, dx_{m+2} \dots dx_n. \qquad \textbf{8.1.7}$$

Note that we use the same notation f for both densities. This is a very common and convenient convention: that, sometimes at least, one uses f to indicate merely 'density of' rather than a particular function, the relevant variables being indicated within the bracket.

Relation **8.1.7** follows from **8.1.1** if we consider an H which is a function only of $X_1, X_2, \ldots, X_m$. The other variables integrate out, and leave us with an m-fold integral in which the modified density **8.1.7** takes the place of the original one. Since this formula is valid for expectations of simple functions of $X_1, X_2, \ldots, X_m$, expression **8.1.7** is by definition the probability density of these variables.

Expression **8.1.7** is sometimes known as the *marginal density* of $X_1, X_2, \ldots, X_m$, although the expression is more usually applied to the single variable densities $f(x_1), f(x_2), \ldots$.

Theorem 8.1.3

If $X_1, X_2, \ldots, X_n$ possess a density and are independent, then

$$f(x_1, x_2, \ldots, x_n) = \prod_1^n f(x_j). \tag{8.1.8}$$

This is the result that one might expect; it follows from the factorization of the indicator function

$$I_{\cap A_j} = \prod_j I_{A_j}, \tag{8.1.9}$$

where A_j is the set $X_j \leqslant x_j$.

Taking expectations of **8.1.9** we obtain, by independence,

$$F(x_1, x_2, \ldots, x_n) = \prod_1^n F(x_j), \tag{8.1.10}$$

and **8.1.8** then follows by differentiation. Note that **8.1.10** is valid, whether or not a density exists.

We have already had examples of densities: the *uniform* or *rectangular* distribution **2.4.7** (see also the exercises of section 2.4), and the *exponential* and *gamma* distributions **4.8.2** and **4.8.4**. The exponential distribution (the continuous analogue of the geometric distribution) holds for the intervals between events in a Poisson process; correspondingly, it is the lifetime distribution for a component whose failure rate is constant with age (exercise 5.1.4).

Exercises

8.1.1   Suppose $X_1, X_2, \ldots, X_n$ are independent r.v.s. having distribution functions $F_j(x_j)$. Show that the distribution function of $\max(X_1, X_2, \ldots, X_n)$ is $\prod_1^n F_j(x)$. Hence show that the minimum of n independent exponential r.v.s is also exponentially distributed. (This result has applications when one considers the failure time of complex electronic or mechanical systems.)

8.1.2 Suppose that $X_1, X_2, \ldots, X_n$ are independently and rectangularly distributed on $(0, L)$. Show that the rth of these r.v.s in order of magnitude has density

$$f(x) = \frac{n!}{(r-1)!(n-r)!} \times \frac{x^{r-1}(L-x)^{n-r}}{L^n} \quad (0 \leqslant x \leqslant L).$$

This is the so-called *beta distribution*. If n and L tend to infinity in such a way that $n/L \to \rho$, then

$$f(x) \to \frac{\rho^r e^{-\rho x} x^{r-1}}{(r-1)!} \quad (x \geqslant 0),$$

which is the *gamma distribution*.

8.1.3 A point source of light is placed at the origin of the (x, y) plane, and is equally likely to release a photon (with straight-line flight path) in any direction. Thus, if θ is the angle between the flight path and the y-axis, θ can be taken as uniformly distributed on $[-\pi, \pi)$.

Consider the distribution of θ conditional on the fact that the photon strikes the line $y = a$ (a positive). Show that θ then has a conditional density which is uniform on $(-\frac{1}{2}\pi, \frac{1}{2}\pi)$. (We have taken pains to indicate whether the ends of the relevant θ interval are open or closed. This is decided, sometimes by convention, and sometimes, as in the second case, by the physics of the situation. It is rarely an important point, since it in any case concerns an event of probability zero.)

8.1.4 *The Buffon needle problem* A needle of length L is dropped onto a floor made of parallel boards of width D. If X is the distance of the centre of the needle from the nearest crack (between floorboards) and θ the angle the needle makes with this crack, then X and θ may be supposed independently and rectangularly distributed over the appropriate ranges. Show that, if $L < D$, the probability that the needle intersects a crack is $2L/\pi D$.

Buffon made an empirical estimate of π from the proportion of times this event occurred in several hundred trials.

8.1.5 A particle is released in a cloud chamber: the time until its first collision and the time until it leaves the chamber are independent exponential variables with means α^{-1} and β^{-1} respectively. Show that the probability that a collision is observed in the chamber is $\alpha/(\alpha + \beta)$.

8.2 Functions of random variables

One very often wishes to consider r.v.s derived from a given set of r.v.s by a functional transformation. We have already considered functions such as $\sum_1^n X_j$ and $\max(X_1, X_2, \ldots, X_n)$, and shall encounter other examples. A natural question to ask is : how do probability densities transform under such a change of variable, if, indeed, the new r.v.s do possess a density at all?

Our guiding principle will be the following. Suppose r.v.s Y_j are derived from r.v.s X_j by a transformation

$$Y_j = y_j(X_1, X_2, \ldots, X_n) \quad (j = 1, 2, \ldots, m) \qquad \textbf{8.2.1}$$

or, in vector terms, $Y = y(X)$, where the $y_j(\ )$ are Borel functions. Suppose also that X has a density $f(x)$, and that the expectation integral can be rewritten

$$E[H(Y)] = \int H[y(x)]f(x)\,dx = \int H(y)g(y)\,dy \qquad \textbf{8.2.2}$$

identically in H. Then $g(y)$ is by definition the density of Y. For, if $H(y)$ is a simple function of y, then $H[y(x)]$ is a Borel function of x, so that **8.2.2** is then valid for a class of functions $H(Y)$ including the simple functions.

Theorem 8.2.1

Suppose that the transformation **8.2.1** *is such that* $Y_1, Y_2, \ldots, Y_m$ *can be complemented by a set of Borel functions*

$$Z_j = z_j(X_1, X_2, \ldots, X_n) \quad (j = 1, 2, \ldots, n - m), \qquad \textbf{8.2.3}$$

where the transformation $X \rightarrow (Y, Z)$ *is one to one, possesses an inverse, and a Jacobian*

$$J(Y, Z) = \frac{\partial(X)}{\partial(Y, Z)}. \qquad \textbf{8.2.4}$$

Then Y has density

$$g(y) = \int f[x(y, z)]J(y, z)\,dz. \qquad \textbf{8.2.5}$$

By the Jacobian **8.2.4** we mean the absolute value of the determinant whose (j, k)th element is $\partial X_j / \partial Y_k$ $(j, k = 1, 2, \ldots, n)$, with Y_k identified with Z_{k-m} for $k > m$. The notation $x(y, z)$ in **8.2.5** indicates that the transformation given by **8.2.1** and **8.2.3** is to be inverted to express x in terms of y and z; the Jacobian **8.2.4** is correspondingly expressed wholly in terms of Y and Z.

The proof follows from the usual formula for a change of variables under a multiple integral (cf. Apostol, 1957, p. 270). Transforming from x to y and z under the integral we have

$$\int H[y(x)]f(x)\,dx = \int\!\!\int H(y)f[x(y, z)]J(y, z)\,dy\,dz, \qquad \textbf{8.2.6}$$

which is of the form **8.2.2** with g determined by **8.2.5**.

If the transformation is simply one between scalar variables y and x then **8.2.5** reduces to

$$g(y) = f[x(y)]\left|\frac{dx}{dy}\right|, \qquad\qquad \textbf{8.2.7}$$

with $|dx/dy|$ expressed as a function of y.

For example, suppose we consider the situation of exercise 8.1.3, and ask for the distribution of the x-coordinate $X = a \tan\theta$ of the point of impact of a photon on the line $y = a$ (conditional on the fact that the photon *does* hit the line). We know that the density of θ is $1/\pi$ if $|\theta| < \frac{1}{2}\pi$, and zero otherwise. Thus, by formula **8.2.7** (but transforming from θ to X rather than from X to Y) we see that the density of X is

$$f(x) = \frac{1}{\pi} \cdot \frac{d\theta}{dx} = \frac{1}{\pi} \cdot \frac{d}{dx} \tan^{-1} \frac{x}{a}$$

$$= \frac{a}{\pi(a^2 + x^2)}, \qquad\qquad \textbf{8.2.8}$$

which is the density of the so-called *Cauchy distribution*.

One of the simplest transformations is the linear one,

$$Y = AX + B. \qquad\qquad \textbf{8.2.9}$$

If this is to be nonsingular then A must be square and A^{-1} must exist. In this case, **8.2.5** will reduce to

$$g(y) = \|A\|^{-1} f[A^{-1}(y - B)], \qquad\qquad \textbf{8.2.10}$$

where $\|A\|$ is the absolute value of the determinant A.

As an important special case, suppose that X is scalar and follows the *standard normal distribution*,

$$f(x) = \frac{1}{\sqrt{(2\pi)}} e^{-\frac{1}{2}x^2} \quad (-\infty < x < \infty), \qquad\qquad \textbf{8.2.11}$$

which will recur repeatedly in our work. We shall see (section 8.5) that X has zero mean and unit variance, so that the r.v.

$$y = \mu + \sigma x \qquad\qquad \textbf{8.2.12}$$

has mean μ and variance σ^2. It then follows from **8.2.10** and **8.2.11** that y has density

$$g(y) = \frac{1}{\sqrt{(2\pi\sigma^2)}} \exp\left[-\frac{1}{2}\left(\frac{y-\mu}{\sigma}\right)^2\right]. \qquad\qquad \textbf{8.2.13}$$

This is the *general normal density*, for a normal r.v. of mean μ and variance σ^2.

155   Functions of random variables

Exercises

8.2.1   Suppose that $Y = h(X)$, where h is a continuous monotonic function. If X and Y have distribution functions $F(x)$ and $G(y)$ respectively, show that $G(y)$ equals $F(x)$ or $1 - F(x-)$, according as h is increasing or decreasing. Show that this result implies **8.2.7** when densities exist.

8.2.2   Suppose the transformation $x \rightarrow y$ is many-to-one, so that there are several x-values yielding the same value of y. Show that, under appropriate regularity conditions, **8.2.7** can then be generalized to

$$g(y) = \sum_j f[x_j(y)] \left| \frac{dx(y)}{dy} \right|_{x = x_j(y)},$$

where the $x_j(y)$ are the roots of $y(x) = y$. Hence show that if velocity X follows distribution **8.2.11**, then kinetic energy $Y = \frac{1}{2}X^2$ has density $e^{-y}/\sqrt{(\pi y)}$ $(y \geq 0)$.

8.2.3   Suppose the position Y of a piston in the cylinder of a car engine is related to the angular position X of the crankshaft by $Y = a \cos X$. If X is equally likely to have any value in the range $[-\pi, \pi)$, show that

$$f(y) = \frac{1}{\pi \sqrt{(a^2 - y^2)}} \quad (-a \leq y \leq a).$$

The U-shaped form of this distribution corresponds to the fact that the piston spends most time near the ends of its stroke, where the motion is slowest.

8.2.4   Suppose that a positive r.v. X has distribution function $F(x)$, and that Y is the rounding error if X is rounded off to the nearest integer below. Show that Y has distribution function

$$\sum_{n=0}^{\infty} [F(n+y) - F(n)] \quad (0 \leq y < 1).$$

8.2.5   Show that if X has a differentiable distribution function $F(x)$, and $Y = F(X)$, then Y is uniformly distributed on $(0, 1)$. This is known as the *probability integral transformation*, and is useful in a number of statistical problems.

8.2.6   Consider the light-emission problem discussed in the text, but in three dimensions, light being radiated uniformly from a point source in all directions. Consider these photons striking a plane distant a from the source. If X is the absolute distance of the point of impact from the foot of the perpendicular from the source to the plane, show that

$$f(x) = \frac{ax}{(a^2 + x^2)^{\frac{3}{2}}} \quad (x \geq 0).$$

8.2.7   Suppose that $X_1, X_2, \ldots, X_n$ are independently distributed, each following the standard normal distribution **8.2.11**. Show that if we consider the linearly

156   Continuous distributions

transformed variables **8.2.9**, then Y has density

$$\frac{|V|^{-\frac{1}{2}}}{(2\pi)^{n/2}}\exp[-\tfrac{1}{2}(Y-\mu)'V^{-1}(Y-\mu)], \qquad \textbf{8.2.14}$$

where $\mu = E(Y)$, and V is the covariance matrix of Y. This is the *multivariate normal distribution*. Note one consequence of this formula: if X_1 and X_2 are jointly normally distributed and are *uncorrelated*, then they are also *independent*.

8.2.8   Continuing the last example, suppose we say that X is a *standard normal vector r.v.* Show that if $Y = AX$, where A is orthogonal, then Y is also a standard normal vector r.v.

   It will already be apparent that one of the functions of r.v.s which occurs most frequently is the *sum*.

Theorem 8.2.2

If $X_1, X_2, \ldots, X_n$ have density $f(x_1, x_2, \ldots, x_n)$, then $S = \sum_{1}^{n} X_j$ has density

$$g(s) = \int \cdots \int f\left(s - \sum_{2}^{n} x_j, x_2, x_3, \ldots, x_n\right) dx_2 \ldots dx_n. \qquad \textbf{8.2.15}$$

This is an immediate consequence of **8.2.5**, if we take $Y = S$ and $Z = (X_2, X_3, \ldots, X_n)$.
   For example, consider the r.v. with density

$$f(x) = \begin{cases} \dfrac{e^{-x}x^{v-1}}{\Gamma(v)} & (x \geq 0), \\ 0 & (x < 0), \end{cases} \qquad \textbf{8.2.16}$$

where v is a positive constant, and the normalizing constant

$$\Gamma(v) = \int_{0}^{\infty} e^{-x}x^{v-1}\, dx \qquad \textbf{8.2.17}$$

is the so-called *gamma function* of v. We have already encountered distribution **8.2.16** for the case of integral v (when $\Gamma(v) = (v-1)!$) in section 4.8, as the density of time lapse until the occurrence of the vth event in a Poisson process (see also exercise 8.1.2). Let us say that a r.v. with density **8.2.16** is a $\Gamma(v)$ variable.

Suppose that X_1 and X_2 are independent, and are respectively $\Gamma(v_1)$ and $\Gamma(v_2)$ variables. By **8.2.15**, the density of the sum $X_1 + X_2$ is

$$g(s) = \frac{e^{-s}}{\Gamma(v_1)\Gamma(v_2)} \int_0^s (s - x_2)^{v_1 - 1} x_2^{v_2 - 1} \, dx_2$$

$$= e^{-s} s^{v_1 + v_2 - 1} \frac{\displaystyle\int_0^1 (1 - u)^{v_1 - 1} u^{v_2 - 1} \, du}{\Gamma(v_1)\Gamma(v_2)} \qquad (s \geqslant 0). \qquad \textbf{8.2.18}$$

The final quotient is a constant, which must necessarily have the value $\Gamma(v_1 + v_2)^{-1}$, as indeed it does. We see from **8.2.18** that S is a $\Gamma(v_1 + v_2)$ variable. The result may obviously be extended: if in the theorem the X_j are independent, and each X_j is a $\Gamma(v_j)$ variable ($j = 1, 2, \ldots, n$), then S is a $\Gamma\left(\sum_1^n v_j\right)$ variable.

This is not too surprising, because we know from section 4.8 that, for integral v, a $\Gamma(v)$ variable is the sum of v independent exponential variables. Here we see that the result extends to non-integral v.

Actually, for sums of independent r.v.s we have a much more powerful technique at our command than the use of the direct formula **8.2.15**; this is the use of *characteristic functions* (section 8.4), the adaptation to the general case of the probability-generating functions we have already employed for the same purpose with integer-valued variables.

Exercises

8.2.9   Suppose that a dynamical system has components of momentum, $X_1, X_2, \ldots, X_n$ which are independent standard normal r.v.s. Show that the kinetic energy $\frac{1}{2} \sum_1^n X_j^2$ is a $\Gamma(\frac{1}{2}n)$ variable (cf. exercise 8.2.2). In another interpretation, $X_1, X_2, \ldots$ are the components of 'miss distance' if one is shooting at a target in n dimensions; $\sqrt{(\sum X_j^2)}$ is then the absolute miss distance.

8.2.10   Show that if X_1 and X_2 are independent r.v.s both following the Cauchy distribution **8.2.8**, then the mean $\frac{1}{2}(X_1 + X_2)$ also follows this distribution. One can prove by the same direct method that the same result holds for the mean $n^{-1} \sum_1^n X_j$ of n such variables, although the result follows much more easily by the methods of section 8.4. Thus the mean incidence point of n independent photons has the same distribution as the incidence point of a single photon, so this is one case where an empirical mean does *not* converge. This lack of convergence is associated with the fact that $E(X)$ does not exist for a Cauchy distribution (see section 9.5).

8.2.11 Show that if X is a $\Gamma(v)$ variable, then

$$E\left(\frac{X}{v} - 1\right)^2 \to 0 \quad \text{as} \quad v \to \infty.$$

Interpret this as mean square convergence of a sample average.

8.2.12 Show that if X_1, X_2 have density $f(x_1, x_2)$, then $Y = X_2/X_1$ has density

$$g(y) = \int f(x, xy)|x| \, dx.$$

8.3 Conditional densities

We have already considered densities conditioned by an event of positive probability (sections 5.1 and 5.4): our aim is now to formulate the obvious result when we have a conditioning field of events of zero probability. In particular, we should like to consider the distribution of a r.v. Y when the conditioning field is that generated from sets $X \leqslant x$, where X is another r.v.

Theorem 8.3.1

Suppose that X and Y are r.v.s (possibly vector) with joint density $f(x, y)$. Then the distribution of Y conditioned by the value of X has density

$$f(y|x) = \frac{f(x, y)}{\displaystyle\int f(x, y) \, dy} = \frac{f(x, y)}{f(x)}. \qquad \textbf{8.3.1}$$

What we must demonstrate is that

$$E^*(K(Y)|X) = \int K(y) f(y|X) \, dy \qquad \textbf{8.3.2}$$

can truly be identified with $E(K(Y)|X)$ for any simple function K; i.e. that E^* satisfies the basic equation **5.3.4**, so that

$$E[H(X)K(Y)] = E[H(X)E^*(K(Y)|X)] \qquad \textbf{8.3.3}$$

for simple functions H, K. But we see from **8.3.1** and **8.3.2** that

$$E[H(X)E^*(K(Y)|X)] = \int\!\!\int H(x) f(x, u) \frac{\displaystyle\int K(y) f(x, y) \, dy}{\displaystyle\int f(x, y) \, dy} \, dx \, du$$

$$= \int\!\!\int H(x) K(y) f(x, y) \, dx \, dy, \qquad \textbf{8.3.4}$$

so that **8.3.3** is indeed valid.

As an example, suppose that X and Y are scalars, jointly normally distributed (see exercise 8.2.7) with zero means, variances V_{XX} and V_{YY}, and covariance V_{XY}. Then by **8.3.1** and formula **8.2.14**,

$$f(y|x) = \left(\frac{V_{XX}}{2\pi(V_{XX}V_{YY} - V_{XY}^2)}\right)^{\frac{1}{2}} \exp\left(-\frac{1}{2}\begin{bmatrix} x \\ y \end{bmatrix}'\begin{bmatrix} V_{XX} & V_{XY} \\ V_{XY} & V_{XX} \end{bmatrix}^{-1}\begin{bmatrix} x \\ y \end{bmatrix} + \frac{x^2}{2V_{XX}}\right)$$

$$= \frac{1}{(2\pi V_{Y|x})^{\frac{1}{2}}} \exp\left(-\frac{(y - \beta x)^2}{2V_{Y|x}}\right), \quad \text{say}, \qquad 8.3.5$$

where

$$\beta = \frac{V_{XY}}{V_{XX}}, \qquad 8.3.6$$

$$V_{Y|X} = V_{YY} - \frac{V_{XY}^2}{V_{XX}}. \qquad 8.3.7$$

That is, the conditional density of Y is again normal, with mean βX and variance $V_{Y|x}$. This has interesting implications. For example, the mean βX must be identifiable with the conditional mean $E(Y|X)$, and we know (Theorem 5.3.1) that $E(Y|X)$ is identifiable with the general least square predictor of Y in terms of X. But in this case $E(Y|X)$ turns out to be linear in X, so it must be identifiable also with the least square *linear* predictor of Y in terms of X. The conditional variance $V_{Y|x}$ is just the mean square prediction error.

The practical relevance of this example is, that it is sometimes quite reasonable to assume variables jointly normally distributed, and we see that in such a case the ideas of linear least square approximation are very natural.

As a second example, suppose we wished to consider a probability density uniform on the surface of the unit sphere in three dimensions. This is a situation which occurs quite frequently: we might, for example, wish to consider a point source in three dimensions which radiates uniformly in all directions, and so gives uniform flux through the surface of a sphere centred on the source. However, the distribution, although simple in concept, is rather difficult to express, because there is no set of coordinates for the surface of a sphere which is natural in the sense that it obviously expresses equivalence of all points on the surface.

Almost the simplest way of going about the problem is the following. Consider the sphere as embedded in a Cartesian space with coordinates X, Y, Z, so that the surface of interest is

$$X^2 + Y^2 + Z^2 = 1. \qquad 8.3.8$$

Suppose that we consider X, Y, Z to be r.v.s with a density $f(x, y, z)$, which is *isotropic* in the sense that

$$f(x, y, z) = g(r) \quad \text{where} \quad r = (x^2 + y^2 + z^2)^{\frac{1}{2}}. \qquad 8.3.9$$

That is, the three-dimensional density f depends only on absolute distance from the origin, r, and not on direction. If we consider the density of X, Y, Z conditional on **8.3.8**, that is on $R = 1$, we shall then be considering just the density we wish: a density uniform on the two-dimensional surface **8.3.8**.

It is simplest to transform to polar coordinates R, θ, ϕ; so that

$X = R\cos\phi,$

$Y = R\sin\phi\cos\theta,$ **8.3.10**

$Z = R\sin\phi\sin\theta,$

with $R \geqslant 0$, $0 \leqslant \phi \leqslant \pi$, $0 \leqslant \theta < 2\pi$. By **8.2.5** and **8.3.9** we see that R, ϕ, θ have density

$$f(r, \phi, \theta) = g(r)r^2|\sin\phi|. \qquad \textbf{8.3.11}$$

It then follows from **8.3.1** that the joint density of ϕ and θ conditional on $R = 1$ is

$$f(\phi, \theta) = \frac{f(r, \phi, \theta)}{\displaystyle\iint f(r, \phi, \theta)\,d\phi\,d\theta}$$

$$= \frac{1}{4\pi}|\sin\phi| \quad (0 \leqslant \phi \leqslant \pi, 0 \leqslant \theta < 2\pi). \qquad \textbf{8.3.12}$$

This expresses the desired density in terms of the angular coordinates ϕ, θ. Naturally, the final density **8.3.12** is independent of the particular function g introduced in **8.3.9**. In fact, assumption **8.3.9** implies that the three r.v.s R, ϕ and θ are statistically independent.

The idea that the conditioning event $R = 1$ is only meaningful when embedded in a field of events (generated from $R \leqslant r$) comes very much to the fore in this example. In effect, it is by this embedding that one *defines* what is meant by 'uniformity' on the surface of a sphere.

Exercise

8.3.1 Show, for the example associated with equation **8.3.5**, that if X and Y have non-zero means, then the conditional distribution of Y is normal with mean $\mu_Y + \beta(X - \mu_X)$ and variance $V_{Y|X}$.

8.4 Characteristic functions

We have already seen (chapter 4, and sections 5.6, 6.3, 6.6, 7.6 and 7.7) how useful the probability-generating function is for the treatment of integral-valued r.v.s, particularly for discussion of sums of such independent r.v.s. However, an attempt to retain the p.g.f. $E(z^X)$ for a general r.v. X is impeded by the fact that z^X is not uniquely defined if X is non-integral. In order to attain uniqueness, it is better to modify the definition to $E(e^{\theta X})$, or, for reasons which will be made clear by Theorem 8.4.1(i), to $E(e^{i\theta X})$.

Thus, if X is a r.v. we shall define

$$\phi(\theta) = E(e^{i\theta X}) \qquad\qquad 8.4.1$$

as the *characteristic function* (c.f.) of X, and we shall occasionally write it as $\phi_X(\theta)$, to emphasize its relation to X.

If X is integral valued, then ϕ is related to the p.g.f. $\Pi(z)$ by

$$\phi(\theta) = \Pi(e^{i\theta}), \qquad\qquad 8.4.2$$

and so differs from it only by a transformation of the argument.

We note a number of elementary properties of c.f.s.

Theorem 8.4.1

(i) $\phi(\theta)$ *exists for* θ *real, when* $|\phi(\theta)| \leqslant 1$, *with* $\phi(0) = 1$.

(ii)   *If* θ *is real then* $\overline{\phi(\theta)} = \phi(-\theta)$.

(iii)   *If* a *and* b *are constants, then*

$$\phi_{a+bX}(\theta) = e^{i\theta a}\phi_X(b\theta).$$

In particular, the c.f. of a r.v. taking the value a *with probability one is* $e^{i\theta a}$.
Assertion (i) follows from the fact that

$$|E(e^{i\theta X})| \leqslant E|e^{i\theta X}| = E(1) = 1. \qquad\qquad 8.4.3$$

The others are immediate consequences of the definition.

However, the key property of a c.f., as it is of a p.g.f., is the following.

Theorem 8.4.2

If X *and* Y *are independent r.v.s, then*

$$\phi_{X+Y}(\theta) = \phi_X(\theta)\phi_Y(\theta). \qquad\qquad 8.4.4$$

This follows from the fact that, for independent r.v.s X and Y,

$$E(e^{i\theta(X+Y)}) = E(e^{i\theta X})E(e^{i\theta Y}). \qquad\qquad 8.4.5$$

Again, the result extends to the sum of several variables.

Another characteristic which is useful, although secondary, is the moment-generating property: by formally expanding the exponential under the expectation sign in **8.4.1** we obtain

$$\phi(\theta) = \sum_{k=0}^{\infty} \frac{(i\theta)^k}{k!} E(X^k), \qquad\qquad 8.4.6$$

$$E(X^k) = i^{-k}\phi^{(k)}(0) \quad (k = 0, 1, 2, \ldots) \qquad\qquad 8.4.7$$

(cf. exercise 4.2.2). However, these results cannot be generally valid, because $E(X^k)$ need not necessarily exist: we shall be more explicit in Theorem 8.4.4.

We may ask whether knowledge of $\phi(\theta)$ for real θ determines the distribution. To a large extent it does, as we shall see in section 11.2. For the moment, we merely note without proof (but see exercise 8.4.9) that, if X has a density $f(x)$, so that

$$\phi(\theta) = \int_{-\infty}^{\infty} e^{i\theta x} f(x)\, dx, \qquad\qquad\qquad\qquad \textbf{8.4.8}$$

then the inverse relation holds, determining f in terms of ϕ:

$$f(x) = \frac{1}{2\pi} \int_{-\infty}^{\infty} e^{-i\theta x} \phi(\theta)\, d\theta. \qquad\qquad\qquad\qquad \textbf{8.4.9}$$

The c.f.s for the standard distributions we have encountered hitherto are most easily presented in a table.

Distribution	Probability or density	Characteristic function
Degenerate	$X = a$ with probability one	$e^{i\theta a}$
Binomial	$P_r = \binom{n}{r} p^r q^{n-r}$	$(pe^{i\theta} + q)^n$
Poisson	$P_r = \dfrac{e^{-\lambda}\lambda^r}{r!}$	$\exp[\lambda(e^{i\theta} - 1)]$
Negative binomial	$P_n = \binom{n-1}{r-1} p^r q^{n-r}$	$\left(\dfrac{pe^{i\theta}}{1 - qe^{i\theta}}\right)^r$
Rectangular	$f(x) = (b-a)^{-1}$ on (a, b)	$\dfrac{e^{i\theta b} - e^{i\theta a}}{i\theta(b - a)}$
Exponential	$f(x) = \rho e^{-\rho x}\ (x \geqslant 0)$	$\left(1 - \dfrac{i\theta}{\rho}\right)^{-1}$
Gamma	$f(x) = \dfrac{\rho^v e^{-\rho x} x^{v-1}}{\Gamma(v)}$	$\left(1 - \dfrac{i\theta}{\rho}\right)^{-v}$
Standard normal	$f(x) = \dfrac{1}{\sqrt{(2\pi)}} e^{-\frac{1}{2}x^2}$	$e^{-\frac{1}{2}\theta^2}$
Normal	$f(x) = \dfrac{1}{\sqrt{(2\pi\sigma^2)}} \exp\left[-\dfrac{1}{2}\left(\dfrac{x-\mu}{\sigma}\right)^2\right]$	$\exp(i\theta\mu - \frac{1}{2}\sigma^2\theta^2)$
Cauchy	$f(x) = \dfrac{a}{\pi(a^2 + x^2)}$	$e^{-a|\theta|}$

Verification of either the formula for ϕ, or its inverse **8.4.9**, is a matter of integration, which we shall leave to the reader. There are some special points, a few of which we shall cover as the need arises.

Exercises

8.4.1 State the analogues of the assertions in Theorem 8.4.1 for a p.g.f.

8.4.2 Note that we have already appealed to the idea of a c.f. when discussing the distribution of total income within a sample (exercise 4.5.5) and of energy in an assembly of molecules (exercises 6.5.2 and 6.5.3).

8.4.3 Note that a $\Gamma(v)$ variable has c.f. $(1-i\theta)^{-v}$; from this and Theorem 8.4.2 the assertion of section 8.2, that a sum of independent $\Gamma(v_1), \Gamma(v_2), \ldots, \Gamma(v_n)$ variables is a $\Gamma\left(\sum_1^n v_j\right)$ variable, follows immediately. At least, this is so if the c.f. really characterizes the distribution: a proviso to be understood also in exercises 8.4.6–8, and to be justified in exercise 8.4.9.

8.4.4 Continuing exercise 7.6.4, show that the first-passage time from state j to state k has c.f.

$$\prod_{r=j}^{k-1}\left(1-\frac{i\theta}{\lambda_r}\right)^{-1}$$

8.4.5 Consider the r.v. $Y(t) = n(t)/E[n(t)]$ for the simple birth–death process with $n(0) = 1$. Show from formula **7.7.8** that if $\lambda > \mu$ this has the limiting $(t \to \infty)$ c.f.

$$\frac{\mu}{\lambda}+\left(1-\frac{\mu}{\lambda}\right)(1-\alpha i\theta)^{-1} \quad \text{where} \quad \alpha = \frac{\lambda}{\lambda-\mu}.$$

Interpret this formula.

8.4.6 Show that a sum of independent normal variables is normal. In particular, if $X_1, X_2, \ldots, X_n$ are independent standard normal variables, then

$$\frac{1}{\sqrt{n}}\sum_1^n X_j \quad \text{and} \quad \frac{\sum c_j X_j}{\sqrt{(\sum c_j^2)}}$$

are standard normal. Here the c_j are constants.

8.4.7 A standard Cauchy variable is one for which $a = 1$ in **8.2.8**. Show that if $X_1, X_2, \ldots, X_n$ are independent standard Cauchy variables, then

$$n^{-1}\sum_1^n X_j \quad \text{and} \quad \frac{\sum c_j X_j}{\sum |c_j|}$$

are standard Cauchy variables.

8.4.8 Consider a $\Gamma(v)$ variable X. Show that in the limit of large v the r.v. $(X-v)/\sqrt{v}$ has c.f. $e^{-\frac{1}{2}\theta^2}$. Interpret.

8.4.9 Show, by integration of the formula **8.4.8** and appeal to the form of c.f. of a normal distribution given in the table, that

$$\frac{1}{2\pi} \int \phi(\theta) \exp(-i\theta x - \tfrac{1}{2}\alpha^2 \theta^2) \, d\theta = \int f(x - \alpha u) h(u) \, du,$$

where $h(u)$ is a standard normal density. By letting α tend to zero we then see that the inversion **8.4.9** must hold, at least at continuity points of f.

The following theorem clears the way for a proper treatment of the moment-generating property.

Theorem 8.4.3

If X is finite with probability one, then $\phi(\theta)$ is uniformly continuous. If $E|X|^j < \infty$ then

$$\phi_j(\theta) = E(X^j e^{i\theta X})$$

is uniformly continuous in θ.
Note that for real θ and θ'

$$|e^{i\theta} - e^{i\theta'}| \leqslant 2,\tag{8.4.10}$$

and also that

$$|e^{i\theta} - e^{i\theta'}| = \left| \int_{\theta'}^{\theta} \frac{de^{i\zeta}}{d\zeta} \, d\zeta \right| \leqslant \left| \int_{\theta'}^{\theta} \left| \frac{de^{i\zeta}}{d\zeta} \right| d\zeta \right| = |\theta - \theta'|.\tag{8.4.11}$$

Thus

$$|\phi(\theta) - \phi(\theta')| \leqslant E(|e^{i\theta X} - e^{i\theta' X}|)$$
$$\leqslant 2P(|X| \geqslant A) + A|\theta - \theta'|,\tag{8.4.12}$$

as we see by using the bound **8.4.10** or **8.4.11** under the expectation, according as $|X| \geqslant A$ or $|X| < A$. In virtue of the finiteness assumption we can choose A so large that $2P(|X| \geqslant A) \leqslant \tfrac{1}{2}\varepsilon$, and then $|\theta - \theta'|$ so small that $A|\theta - \theta'| \leqslant \tfrac{1}{2}\varepsilon$, and so obtain $|\phi(\theta) - \phi(\theta')| \leqslant \varepsilon$, where ε is an arbitrarily small positive quantity. The first assertion is thus established, and the second is proved analogously.

Theorem 8.4.4 (Limited expansion theorem)

If $E(|X|^j) < \infty$ for a given integer j then

$$\phi(\theta) = \sum_{k=0}^{j} \frac{(i\theta)^k}{k!} E(X^k) + o(\theta^j)\tag{8.4.13}$$

*and **8.4.7** holds for $k = 0, 1, 2, \ldots, j$.*

We have the limited Taylor expansion

$$e^{i\theta X} = \sum_{k=0}^{j} \frac{(i\theta X)^k}{k!} + \frac{(i\theta)^j}{(j-1)!} \int_0^1 (e^{i\theta Xt}X^j - X^j)(1-t)^{j-1}\,dt. \qquad \textbf{8.4.14}$$

Taking expectations, we find

$$\phi(\theta) = \sum_{k=0}^{j} \frac{(i\theta)^k}{k!} E(X^k) + \frac{(i\theta)^j}{(j-1)!} \int_0^1 [\phi_j(\theta t) - \phi_j(0)](1-t)^{j-1}\,dt. \qquad \textbf{8.4.15}$$

But, by theorem 8.4.3, $\phi_j(\theta)$ is continuous, so that the integral in the remainder term of **8.4.15** tends to zero with θ. That is, the remainder as a whole is of smaller order than θ^j, as asserted in **8.4.13**. The second assertion of theorem 8.4.4 follows from **8.4.13**.

One can ask whether the converse result holds: that the existence of $\phi^{(j)}(0)$ implies the existence of $E(X^j)$. The statement is true for j even (see section 11.2) but, surprisingly, not for j odd.

Exercises

8.4.10   Consider a sum $S = \sum_1^n X_j$ where the X_j are independent r.v.s with common c.f. $\phi(\theta)$, and n is also a r.v., independent of the X_j, with p.g.f. $\Pi(z)$. Show that S has c.f. $\Pi[\phi(\theta)]$, and hence that

$E(S) = E(n)E(X),$

$\mathrm{var}(S) = E(n)\,\mathrm{var}(X) + \mathrm{var}(n)[E(X)]^2.$

Thus S might be the total claim paid in a year by an insurance company (a random number of random claims) or the total electric charge in a region carried by molecules of a Poisson process (a random number of variable charges).

8.4.11   Consider the renewal problem if lifetimes may be continuously distributed. Show that under the assumptions of section 6.3 (independent lifetimes with fixed c.f. $\phi(\theta)$, and an initial renewal at $t = 0$), we have

$$\int_0^\infty e^{i\theta t}\,dM(t) = \frac{1}{1 - \phi(\theta)},$$

where $M(t)$ is the expected number of renewals in $[0, t]$, that at $t = 0$ being counted. (Actually, the formula is valid only for θ such that $|e^{i\theta}| < 1$, i.e. $\mathrm{Im}(\theta) > 0$, just as **6.3.5** is valid only for $|z| < 1$.)

8.4.12   Consider the *stochastic difference equation*

$X_j = \rho X_{j-1} + \varepsilon_j,$

where the ε_j are independent r.v.s with c.f. $\phi(\theta)$. Show that if the equation holds back to $j = -\infty$, with $X_{-\infty}$ bounded, and $|\rho| < 1$, then X_j has c.f. $\prod\limits_0^\infty \phi(\rho^k\theta)$.

8.5 The normal distribution; normal convergence

We introduced the normal distribution in formula **8.2.11**, but as yet have found no physical motivation for it, although we have noted a number of significant properties: the multivariate generalization and the orthogonality–independence implication of exercise 8.2.7; the form of the conditional density **8.3.5** and its relation with least square linear approximation; the self-conjugacy of the density (i.e. the fact that $f(x)$ and $\phi(\theta)$ are of the same functional form); the 'closure' property under summation of exercise 8.4.6. In fact, the distribution is not one that arises immediately in physical contexts; nevertheless, it is possibly the most important single distribution in probability theory. The principal reason is its limit property, presaged in exercise 8.4.6, and to be made explicit in Theorems 8.5.1 and 8.5.2.

The normalizing factor $(2\pi)^{-\frac{1}{2}}$ in the density **8.2.11** is correct; see exercise 8.5.1 for an elementary proof. All moments $\mu_j = E(X^j)$ exist. A partial integration yields the recursion

$$\mu_j = (j-1)\mu_{j-2}, \qquad\qquad\qquad\qquad \textbf{8.5.1}$$

whence $\quad \mu_{2j} = \dfrac{(2j)!}{2^j(j!)}: \qquad\qquad\qquad\qquad \textbf{8.5.2}$

the moments of odd order being zero, by symmetry. Thus, for the standard density **8.2.11**, X has zero mean and unit variance.

A very convenient terminology is to say that a r.v. with the general normal density **8.2.13** is $N(\mu, \sigma^2)$. This carries over to the vector case: we say that a vector r.v. with density **8.2.14** is $N(\mu, V)$.

As mentioned, the key characteristic of the normal distribution is its limit property. Suppose we consider a sum $S_n = \sum\limits_1^n X_j$ of independent r.v.s, all possessing a variance, and standardize this to zero mean and unit variance:

$$u_n = \frac{S_n - E(S_n)}{\sqrt{[\text{var}(S_n)]}}. \qquad\qquad\qquad\qquad \textbf{8.5.3}$$

Under wide conditions, u_n has in the limit $n \to \infty$ a normal distribution, in a sense to be explained. Let us first establish the fact that, if a limit distribution exists, it must be normal. We thus postulate the existence of a limiting distribution, and shall say that u_n is in the limit L-distributed.

Now, suppose that $\xi_1, \xi_2, \ldots, \xi_n$ are independent L-distributed variables, so that each is representable as the 'limit' of a standardized sum such as

8.5.3. Since

$$\zeta = \frac{1}{\sqrt{n}} \sum_1^n \xi_j \qquad\qquad \textbf{8.5.4}$$

is also so representable, it is natural to demand that ζ should also be L-distributed, or else the limit property is hardly a stable and useful one. This requirement of stability would follow as a consequence if the limit law were known to be unique, but to demand stability is less than to assume uniqueness.

We thus take this stability or *closure property* as a partial characterization of an L-distribution: that if independent r.v.s $\xi_1, \xi_2, \ldots, \xi_n$ are L-distributed, then so is $\sum_1^n \xi_j/\sqrt{n}$, for any integral n. In fact, the property provides a complete characterization, since there is only one distribution possessing it.

Theorem 8.5.1

The only distribution (of unit variance) possessing the closure property just defined is that with c.f. $e^{-\frac{1}{2}\theta^2}$, the normal distribution.

In terms of the c.f. $\phi(\theta)$ of an L-distributed variable, the closure property amounts to the requirement

$$\phi(\theta) = \left[\phi\left(\frac{\theta}{\sqrt{n}} \right) \right]^n \qquad (n = 1, 2, \ldots) \qquad\qquad \textbf{8.5.5}$$

(Theorems 8.4.1(iii), 8.4.2).

It is plain that $e^{-\frac{1}{2}\theta^2}$ has this property. However, the theorem provides more than this simple verification: it deduces the normal distribution as the only distribution having the closure property. We have not yet 'deduced' the normal distribution in any sense, but in this way it now makes its inevitable entrance on the scene.

It is more convenient to work in terms of $\psi = \log \phi$ rather than ϕ; we have then

$$\psi(\theta) = n\psi\left(\frac{\theta}{\sqrt{n}} \right) = m\psi\left(\frac{\theta}{\sqrt{m}} \right), \qquad\qquad \textbf{8.5.6}$$

for any integral m, n. A change of θ-scale in **8.5.6** then shows that

$$\psi(\theta) = \alpha\psi\left(\frac{\theta}{\sqrt{\alpha}} \right), \qquad\qquad \textbf{8.5.7}$$

for $\alpha = m/n$. Since ψ is continuous, by Theorem 8.4.3, and the rationals are dense, equation **8.5.7** must hold for any positive α. Choosing $\sqrt{\alpha} = |\theta|$, we find then from **8.5.7** that

$$\psi(\theta) = \begin{cases} \psi(1)\theta^2 & (\theta \geqslant 0), \\ \psi(-1)\theta^2 & (\theta \leqslant 0). \end{cases} \qquad\qquad \textbf{8.5.8}$$

However, since the r.v. possesses a variance, which is in fact unity, we see from Theorem 8.4.4 that $-\phi''(0)$ exists, and equals unity. This and equation **8.5.8** imply that $\psi(1) = \psi(-1) = -\frac{1}{2}$; whence $\psi(\theta) = -\frac{1}{2}\theta^2$, which was to be proved.

We have thus established the fact that if a standardized sum u_n has a stable limit distribution, this must be the normal distribution; we should now determine conditions for its actual convergence. More specifically, we should like to find conditions which ensure that

$$E[H(u_n)] \rightarrow E[H(u)], \qquad\qquad 8.5.9$$

where u is $N(0, 1)$, for some useful class of functions H. We can obtain results for the functions $H(u) = e^{i\theta u}$ (θ real) fairly easily.

Theorem 8.5.2 (the central limit theorem)

Suppose that the r.v.s X_j are independently and identically distributed, with $E(X^2) < \infty$. Then u_n converges to normality in that

$$E(e^{i\theta u_n}) \rightarrow e^{-\frac{1}{2}\theta^2}, \qquad\qquad 8.5.10$$

for fixed real θ.

For, by the limited expansion theorem (Theorem 8.4.4), the standardized variable

$$\frac{X - E(X)}{\sqrt{[\mathrm{var}(X)]}}$$

will have c.f.

$$\phi(\theta) = 1 - \tfrac{1}{2}\theta^2 + o(\theta^2) = \exp[-\tfrac{1}{2}\theta^2 + o(\theta^2)], \qquad\qquad 8.5.11$$

so that u_n will have c.f.

$$\left[\phi\left(\frac{\theta}{\sqrt{n}}\right)\right]^n = \exp\left[-\tfrac{1}{2}\theta^2 + n\,o\!\left(\frac{\theta^2}{n}\right)\right] \rightarrow e^{-\frac{1}{2}\theta^2}. \qquad\qquad 8.5.12$$

The conditions on the distribution and independence of the X_j can be relaxed very considerably, but we shall not pursue this point further. The validity of **8.5.9** for more general functions H can be proved either from **8.5.10** (see section 11.2), or by the direct methods of the next section.

A final practical point: the distribution function of an $N(0, 1)$ variable

$$\Phi(x) = \frac{1}{\sqrt{(2\pi)}} \int_{-\infty}^{x} e^{-\frac{1}{2}u^2}\, du \qquad\qquad 8.5.13$$

is a non-elementary function known as the *normal integral*. It is important, not only in probability, but also in diffusion theory, and is extensively tabulated.

Exercises

8.5.1   Show by a polar transformation under the integral that

$$\int\int_{-\infty}^{\infty} \exp[-\tfrac{1}{2}(x^2 + y^2)]\, dx\, dy = 2\pi.$$

8.5.2   Suppose that R is $b(n, p)$. Give a direct proof of normal convergence by showing that for a fixed value of

$$u = \frac{r - np}{\sqrt{(npq)}}$$

we have

$$P(R = r) \to k\, e^{-\frac{1}{2}u^2} \quad (k \text{ constant in } u)$$

as $n \to \infty$. (Use the Stirling approximation $n! \simeq \sqrt{(2\pi)}e^{-n}n^{n+\frac{1}{2}}$.) Why is the constant k not equal to $(2\pi)^{-\frac{1}{2}}$? This asymptotic result, and the tables of Φ, enable one to evaluate quantities such as $P(R > k)$ with good approximation, which leads to convenient statistical tests of whether a proposed value of p can be regarded as concordant with an observed value of R.

8.5.3   Suppose that $X_1, X_2, \ldots, X_n$ are independently distributed, with a continuous density which is a function of $\sum_1^n X_j^2$ alone. Show that they are normally distributed with zero mean and a common variance. (This is Maxwell's characterization, based on the facts that the Cartesian velocity components of a molecule in a perfect gas can be proved independent, and the gas may be assumed isotropic (rotationally invariant) in its properties.)

8.5.4   The limit property of the normal distribution was closely related to the closure property: that is, if $X_1, X_2, \ldots, X_n$ are independent $N(0, 1)$, then $n^{-\frac{1}{2}} \sum_1^n X_j$ is also $N(0, 1)$. But the Cauchy distribution shows a similar closure property: see exercise 8.4.7.

Adapt the argument of Theorem 8.5.1 to show that, if a sequence $\{B_n\}$ exists such that $B_n^{-1} \sum_1^n X_j$ has the same distribution as the individual (independent) X_j, then this distribution must have a c.f. of the form $\exp(-c|\theta|^\gamma)$, where c may take different values for $\theta > 0$ and $\theta < 0$. This is (roughly) the class of *stable laws*: the Cauchy and normal laws correspond to $\gamma = 1, 2$. However, values of γ outside the range $[0, 2]$ are inadmissible, since the function is then not a c.f., while for $\gamma < 2$ the law is one for which the r.v. does not have a variance. Thus the normal law has a unique character.

8.5.5   Results for the summation operation $S_n = \sum_1^n X_j$ often have an analogue for the maximization operation

$$T_n = \max(X_1, X_2, \ldots, X_n).$$

Suppose that sequences of standardizing constants A_n and B_n exist such that

$$u_n = \frac{T_n - A_n}{B_n}$$

has a stable limit distribution function $e^{G(u)}$, if the X_j are independently and identically distributed. Show that there is then a closure relationship

$$n G\left(\frac{u - A_n}{B_n}\right) = G(u).$$

Hence show that, under suitable regularity conditions, G must be of the form $(\alpha + \beta u)^\gamma$: this includes limit forms such as $\delta e^{\varepsilon u}$. These are the *extreme value distributions*, with a theory analogous to that of the stable laws. They are useful for the description of the distribution of extremes in large sets of r.v.s, not necessarily independent (e.g. the distribution of peak floods over a long period of time; the failure probability of a structure with many parts, etc.).

8.5.6   Adapt the proof of Theorem 8.5.2 to show that

$$E\left[\exp\left(\frac{i\theta S_n}{n}\right)\right] \to e^{i\theta\mu}$$

if the X_j are independently and identically distributed, and possess a mean value μ. (Compare with section 6.2.)

8.6   A direct proof of normal convergence

It will be plain from the last section that the characteristic function is a natural tool for this type of problem. On the other hand, it does have the disadvantage mentioned at the end of section 8.5: having once proved (Theorem 8.5.2) the validity of the convergence **8.5.9** for functions $H(u) = e^{i\theta u}$, how do we extend the result to other functions?

This extension can be an untidy and even a difficult analytic exercise. We shall face up to it in section 11.2 and prove, for example, that the convergence of characteristic functions implies the convergence of the expectation of any bounded continuous function. However, in this section we shall consider the proof of **8.5.9** by direct methods, using relation **6.2.12** to replace the key c.f. property of Theorem 8.4.2.

Note that there may indeed be functions $H(u)$ for which the convergence **8.5.9** does not hold. For example, suppose $E(X^r)$ is infinite, for some $r > 2$, where X is a typical summand of **8.5.3**. Then convergence cannot hold for $H(u) = u^r$, because in this case $E(u^r)$ is finite but $E(u_n^r)$ infinite. Again, if X does not have a density then neither does u_n, whereas u does. In such a case, convergence would not hold if H were a Dirac delta function, for example.

Theorem 8.6.1

Suppose that H is a function whose second derivative obeys the uniform continuity condition

$$|H''(u+s) - H''(u)| \leqslant K|s|^\alpha. \tag{8.6.1}$$

Then, under the conditions of Theorem 8.5.2,

$$|E[H(u_n)] - E[H(u)]| \leqslant \frac{Kn^{-\frac{1}{2}\alpha}}{(\alpha+1)(\alpha+2)}[E(|X'|^{\alpha+2}) + E(|u|^{\alpha+2})], \tag{8.6.2}$$

where $\quad X' = \dfrac{X - E(X)}{\sqrt{[\mathrm{var}(X)]}}.$ \hfill 8.6.3

Thus the convergence **8.5.9** will hold if **8.6.1** and the inequality

$$E(|X|^{\alpha+2}) < \infty \tag{8.6.4}$$

hold for some positive α. A supplementary argument (see the comment associated with equation **6.2.9**) extends this conclusion to a considerably wider class of functions.

Statement **8.6.2** is reminiscent of the version of the law of large numbers proved in section 6.2, and follows quickly by the methods used in that section.

Replace the r.v.s X and Y in **6.2.12** by $X'/\sqrt{n}$ and $u/\sqrt{n}$ respectively. We can write the consequent inequality in the form

$$\|E[H(t+u_n)] - E[H(t+u)]\| \leqslant n\left\|E\left[H\left(t + \frac{X'}{\sqrt{n}}\right)\right] - E\left[H\left(t + \frac{u}{\sqrt{n}}\right)\right]\right\|. \tag{8.6.5}$$

Here we have appealed to the fact that $n^{-\frac{1}{2}}\sum_1^n u(j)$ has the same distribution as u, if the $u(j)$ are independent and $N(0,1)$. This closure property is the only explicit feature of the normal distribution that the proof requires. We have thus

$$|E[H(u_n)] - E[H(u)]| \leqslant n\left\|E\left[H\left(t + \frac{X'}{\sqrt{n}}\right)\right] - E\left[H\left(t + \frac{u}{\sqrt{n}}\right)\right]\right\|$$

$$= \Delta_n, \quad \text{say}. \tag{8.6.6}$$

By taking the expectation of the partial Taylor expansion

$$H\left(t + \frac{u}{\sqrt{n}}\right) = H(t) + \frac{u}{\sqrt{n}}H'(t) + \frac{u^2}{2n}H''(t) + R(u) \tag{8.6.7}$$

we find that $\quad E\left[H\left(t + \dfrac{u}{\sqrt{n}}\right)\right] = H(t) + \dfrac{H''(t)}{2n} + E[R(u)],$ \hfill 8.6.8

where $R(u) = \int\limits_{0}^{u/\sqrt{n}} \left(\dfrac{u}{\sqrt{n}} - s \right) [H''(t+s) - H''(t)]\, ds,$ **8.6.9**

so that $\Delta_n = n \| E[R(X')] - E[R(u)] \|.$ **8.6.10**

This quantity will tend to zero with increasing n under a variety of circumstances. In particular, if we make the assumption **8.6.1**, then expression **8.6.10** is easily found to be bounded above by the right-hand member of **8.6.2**. Conditions **8.6.1** and **8.6.4** are thus sufficient for the validity of $E[H(u_n)] \to E[H(u)]$.

Exercise

8.6.1   Suppose that $E(X^r)$ exists, for some $r > 2$. Find an adaptation of the theorem of this section which establishes convergence **8.5.9** for $H(u) = u^r$, and for $H(u) = |u|^r$.

9 Convergence of Random Sequences

9.1 Characterization of convergence

Probability theory is founded on an empirical limit concept, and its most characteristic conclusions take the form of limit theorems. Thus, a sequence of r.v.s $\{X_n\}$ which one suspects has some kind of limit property for large n is a familiar object. For example, we have repeatedly considered (sections 4.2, 6.1 and 6.2) the convergence of a sample average $\bar{X}_n$ to the expected value μ, and of the standardized sums u_n to normality (sections 8.5 and 8.6). Any infinite sum of r.v.s which we encounter should be construed as a limit (in some sense) of a finite sum: consider, for instance, the sum $\sum_0^\infty R_n z^n$ of section 6.3, or the formal solution

$$X_j = \sum_0^\infty \rho^k \varepsilon_{j-k} \qquad\qquad \textbf{9.1.1}$$

of the difference equation in exercise 8.4.12. Yet another example of a sequence which we hope converges is a sequence of expectations conditional on an increasing field.

The question of convergence of a sequence of r.v.s $\{X_n\}$ is rather less straightforward than that of a sequence of constants $\{a_n\}$. One understands quite clearly what is meant by convergence of $\{a_n\}$: that there exists a constant a such that $a_n - a \to 0$; or more precisely, that for any positive ε there exists a number $n(\varepsilon)$ such that $|a_n - a| < \varepsilon$ for all n greater than $n(\varepsilon)$. A necessary and sufficient condition for such convergence is that the sequence should be *mutually convergent*,

i.e. that $\quad |a_m - a_n| \to 0$

as m and n tend to infinity independently. The usefulness of this second criterion is that it does not suppose knowledge of the limit value a. A sequence having the mutual convergence property is sometimes termed a *Cauchy sequence*.

However, when it comes to a sequence of r.v.s $\{X_n\}$ (which is really a sequence $\{X_n(\omega)\}$ of functions on a general space Ω) then one can conceive of many types of convergence. One might first make the requirement of *pointwise convergence*: that $X_n(\omega)$ should tend to $X(\omega)$ for any given ω as n increases.

That is, $X_n \to X$, whatever the outcome of an experiment. However, convergence in such a strong sense is almost completely uncharacteristic of probabilistic models. For example, the proportion of heads in n tosses of a fair coin does *not* converge to $\frac{1}{2}$ for *all* sequences, although the infinite sequences for which it does not have *zero* probability (see Theorem 9.5.1, the strong law of large numbers).

So already it is natural to introduce a weaker concept of convergence; that $X_n(\omega)$ should converge to $X(\omega)$ for all ω except for an ω-set of zero probability (*almost certain convergence*). However, in some physical situations even this is too strong a concept, for there are cases where one can establish only the convergence of certain averages rather than of the sequence of r.v.s itself.

Thus, we might weaken our requirements to the point that we demand that

$$E[H(X_n)] \to E[H(X)] \qquad\qquad\qquad\qquad \textbf{9.1.2}$$

for all functions H of some suitable class (which is what we did in sections 6.2 and 8.6). But then we are really only saying that the distribution of X_n approximates, at least partially, to that of X, and not that X_n itself approximates to X. (The two concepts coincide only in the case where X is constant.) To ensure virtual identity of X and 'lim X_n' one would have to require something like

$$E[H(X_n - X)] \to H(0) \qquad\qquad\qquad\qquad \textbf{9.1.3}$$

for a suitable class of functions H. A generalization of this would be to require that

$$E[H(X_n, Y)] \to E[H(X, Y)], \qquad\qquad\qquad\qquad \textbf{9.1.4}$$

where Y is another r.v. (possibly vector, which may or may not have X as one of its components). We thus ensure that 'lim X_n' bears to some extent the same relation to the reference r.v. Y as X itself does.

This does not exhaust the possibilities; one might make convergence depend on the whole *tail* of the sequence:

$$E[H(X_n, X_{n+1}, \dots ; Y)] \to E[H(X, X, \dots ; Y)] \qquad\qquad\qquad \textbf{9.1.5}$$

for suitable H. Almost certain convergence constitutes just such a demand.

There are obviously many possibilities, and by playing with them one can construct an extensive and somewhat sterile theory. In a given physical situation one must be guided by operational considerations: what kind of convergence is it natural to demand?

Exercise

9.1.1   Determine the functions H which are invoked for (i) mean square convergence (see section 6.1), (ii) convergence in probability, and (iii) almost certain convergence.

9.2 Types of convergence

Let us list some of the types of convergence in standard use.

If the limit relation **9.1.2** holds for any bounded continuous H, then $\{X_n\}$ is said to converge to X *weakly in distribution*. As we mentioned there, this does not imply convergence of X_n to X itself in any sense, except in the case where X has a degenerate distribution, and is essentially constant.

Suppose we consider relation **9.1.3** with $H(u) = |u|^r$ where r is a prescribed positive index. We are thus requiring that

$$E(|X_n - X|^r) \to 0. \qquad \textbf{9.2.1}$$

This is *convergence in the r-th mean, or L_r-convergence*, in brief, and we write

$$X_r \overset{L_r}{\to} X. \qquad \textbf{9.2.2}$$

The case $r = 2$ (*mean square convergence*, or m.s. convergence) which we have already encountered in section 6.1, is particularly important and natural in many situations: for this special case we write

$$X_n \overset{\text{m.s.}}{\longrightarrow} X. \qquad \textbf{9.2.3}$$

Suppose we consider a relation **9.1.3** with H the indicator function of an interval $(-\varepsilon, \varepsilon)$; that is, suppose that

$$P(|X_n - X| < \varepsilon) \to 1. \qquad \textbf{9.2.4}$$

If **9.2.4** can be established for arbitrarily small fixed ε, that is, if for any prescribed positive ε and δ, one can find a number $n(\varepsilon, \delta)$ such that

$$P(|X_n - X| < \varepsilon) > 1 - \delta \qquad \textbf{9.2.5}$$

for all $n > n(\varepsilon, \delta)$, then $\{X_n\}$ is said to *converge to X in probability*, or *P-converge*. We write

$$X_n \overset{P}{\to} X. \qquad \textbf{9.2.6}$$

An appeal to Markov's inequality (cf. section 6.1) shows that **9.2.2** implies **9.2.6**: L_r-convergence implies P-convergence.

Finally, we obtain a special type of the tail-convergence **9.1.5** by setting

$$H(X_n, X_{n+1}, \ldots ; Y) = \begin{cases} 1 \ (|X_m - X| < \varepsilon \text{ for } m \geqslant n), \\ 0 \ (\text{otherwise}), \end{cases} \qquad \textbf{9.2.7}$$

and so requiring that

$$P(|X_m - X| < \varepsilon; m \geqslant n) \to 1 \qquad \textbf{9.2.8}$$

as $n \to \infty$. If **9.2.8** is valid for arbitrarily small positive ε then we say that $\{X_n\}$ converges *almost certainly* (or is a.c. convergent) to X, and write

$$X_n \overset{\text{a.c.}}{\longrightarrow} X. \qquad \textbf{9.2.9}$$

This concept is identical with the almost certain convergence mentioned in the previous section, because what **9.2.8** states is that the event 'X_n converges to X' (in the conventional sense) has probability one.

The types of convergence **9.2.6** and **9.2.9** are sometimes referred to as *weak* and *strong convergence in probability* respectively. This is quite a good nomenclature, but is perhaps best avoided, because of possible confusion with the concept of weak convergence in distribution.

In general, there are a number of implications between the various criteria, summarized by the arrows of the following diagram.

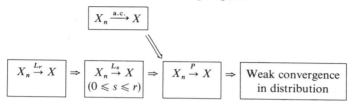

We leave verification of these implications as an exercise to the reader: some hints are given in exercises 9.2.1 and 2. It is a rather less direct matter to prove that there is in fact no implication where none is indicated on the diagram: appropriate counterexamples are indicated in exercises 9.2.3, 4, 5 and 8.

Exercises

9.2.1 Use the fact (exercise 2.7.10) that $\left[E(|X|^r)\right]^{1/r}$ is an increasing function of r to establish the first implication in the bottom row of the diagram.

9.2.2 Use Chebichev's inequality to establish the second.

9.2.3 Neither P-convergence nor L_r-convergence imply a.c. convergence. Consider a sequence of independent variables $\{X_n\}$, for which

$$P(X_n = 0) = 1 - \frac{1}{n} \qquad P(X_n = 1) = \frac{1}{n}.$$

9.2.4 P-convergence does not imply L_r-convergence. Consider an independent sequence with

$$P(X_n = 0) = 1 - \frac{1}{n} \qquad P(X_n = n^{2/r}) = \frac{1}{n}.$$

9.2.5 A.C. convergence does not imply L_r-convergence. Consider $X_n = a_n Y$, where $\{a_n\}$ is a convergent sequence of constants, and $E(|Y|^r) = \infty$.

9.2.6 Not even pointwise convergence implies corresponding convergence of expectations. Consider a sample space $\xi > 0$ with $X_n = n^2 \xi e^{-n\xi}$ and ξ exponentially distributed. Then $X_n \to 0$ pointwise, but $E(X_n) \to 1$.

9.2.7   Show that a m.s. convergent series is mutually m.s. convergent, in that

$$X_m - X_n \xrightarrow{\text{m.s.}} 0.$$

9.2.8   Let $u_n = n^{-\frac{1}{2}} \sum_1^n X_j,$

where the X_j are independent identically distributed and standardized variables, so that, by the results of section 8.5, u_n tends in distribution to normality. Show that it is, however, not m.s. convergent.

9.2.9   Show that a sequence $\{X_n\}$ is a.c. convergent iff it is a.c. mutually convergent,

i.e. $X_m - X_n \xrightarrow{\text{a.c.}} 0.$

9.2.10   Let $I_\varepsilon(x)$ be the indicator function of the x-interval $(-\varepsilon, \varepsilon)$. Note that $\lim_{\varepsilon \downarrow 0} I_\varepsilon(x)$ and $\lim_{r \downarrow 0} |x|^r$ have the same limit,

i.e. $H(x) = \begin{cases} 0 & (x = 0), \\ 1 & (x \neq 0). \end{cases}$

From this point of view, one can formally identify P-convergence with L_0-convergence.

9.3   Some consequences

In this section we shall consider some useful simple conditions which ensure various types of convergence, or which permit one to make some of the reverse implications not included in the diagram of the previous section.

Let us first follow up the helpful fact that, since the event 'mutual convergence' (in the conventional sense) is equivalent to the event 'convergence', then a.c. mutual convergence is equivalent to a.c. convergence.

Theorem 9.3.1

If $\sum P(|X_{n+1} - X_n| \geqslant \varepsilon_n) < \infty$, *where* $\sum \varepsilon_n$ *is a convergent sum of positive terms, then* $\{X_n\}$ *is a.c. convergent.*

For, suppose $\delta_n = \sum_n^\infty \varepsilon_j$. Then

$$P(|X_j - X_k| < \delta_n; j, k \geqslant n) \geqslant P(|X_{j+1} - X_j| < \varepsilon_j; j \geqslant n)$$

$$\geqslant 1 - \sum_n^\infty P(|X_{j+1} - X_j| \geqslant \varepsilon_j) \to 1. \qquad \textbf{9.3.1}$$

Thus $\{X_n\}$ is a.c. mutually convergent and the result follows. Note that the second step is an application of Boole's inequality **3.4.23**.

Theorem 9.3.2

If $\{X_n\}$ is L_r mutually convergent then it contains a subsequence $\{X_n'\}$ which is a.c. convergent.

Since

$$E(|X_m - X_n|^r) \to 0 \qquad\qquad\qquad 9.3.2$$

we can pick out a subsequence $\{X_n'\}$ for which

$$E(|X_{n+1}' - X_n'|^r) \leq \eta_n, \qquad\qquad\qquad 9.3.3$$

where η_n tends to zero sufficiently rapidly with increasing n that $\sum (\eta_n/\varepsilon_n^r) < \infty$. Here $\{\varepsilon_n\}$ is the sequence of the previous theorem. An application of Markov's inequality then yields

$$\sum P(|X_{n+1}' - X_n'| \geq \varepsilon_n) \leq \sum_n \frac{\eta_n}{\varepsilon_n^r} < \infty, \qquad\qquad 9.3.4$$

so that $\{X_n'\}$ is a.c. convergent, by Theorem 9.3.1.

For example, consider the partial sum

$$X_n = \sum_0^n R_t z^t,$$

where R_t is the number of renewals at time t, considered in section 6.3. We have

$$E(|X_{n+1} - X_n|) = |z|^{n+1} E(R_{n+1}) \leq \frac{|z|^{n+1}}{1 - p_0},$$

(see exercise 6.3.2). Thus, if $|z| < 1$, we have

$$\sum P(|X_{n+1} - X_n| \geq |z|^{\frac{1}{2}n}) \leq \frac{\sum |z|^{\frac{1}{2}n}}{1 - p_0} < \infty$$

so that $\{X_n\}$ is a.c. convergent We can thus attach a meaning to the infinite sum $\sum_0^\infty R_n z^n$.

A final result, which we give for interest, is essentially a probabilistic version of the dominated convergence theorem (exercise 11.1.4) but framed under much weaker assumptions.

Theorem 9.3.3

Suppose that $X_n \xrightarrow{P} X$ and $Y \leq X_n \leq Z$ for all n, where $E(|Y|)$ and $E(|Z|)$ are both finite. Then $X_n \xrightarrow{L_1} X$, and $E(X) = \lim E(X_n)$.

179   Some consequences

We have $0 \leqslant E(Z-Y) < \infty$. Let A_n denote the event $|X - X_n| > \varepsilon$, so that $P(A_n) \to 0$. We have

$$E(|X - X_n|) = P(\bar{A}_n)E[(|X - X_n|)\,|\,\bar{A}_n] + P(A_n)E[(|X - X_n|)\,|\,A_n]$$

$$\leqslant \varepsilon + P(A_n)E(Z - Y) \to \varepsilon.$$

Since ε is arbitrarily small, the first result is proved. The second follows from

$$|E(X) - E(X_n)| \leqslant E(|X - X_n|).$$

Exercises

9.3.1 Show that P-convergence implies P mutual convergence, and that a P mutually convergent sequence contains an a.c. convergent subsequence.

9.3.2 Suppose that $\sum E(X_n^2) < \infty$. Show that $X_n \xrightarrow{\text{a.c.}} 0$.

9.3.3 Find a version of Theorem 9.3.3 establishing $X_n \xrightarrow{L_r} X$.

9.4 **Kolmogorov's inequality, and refinements of it**

We have already considered the behaviour of partial sums $S_n = \sum_1^n X_j$ from the point of view of convergence of mean values $\bar{X}_n$ (sections 6.1 and 6.2) and normal convergence of standardized sums u_n (sections 8.5 and 8.6). In this section we shall derive inequalities which lead us to some of the stronger results of classic probability theory.

It will be a convenience, and no loss of generality, if we assume henceforth that the X_j have zero mean. Any result we obtain on this assumption can be rephrased for the general case by setting $X_j - \mu$ wherever X_j occurs.

Theorem 9.4.1 (*Kolmogorov's inequality*)

Suppose that

$$E(X_j\,|\,X_1, X_2, \ldots, X_{j-1}) = 0 \quad (j = 1, 2, \ldots). \qquad \textbf{9.4.1}$$

Then

$$P(|S_j| \leqslant u : j = 1, 2, \ldots, n) \geqslant 1 - \frac{E(S_n^2)}{u^2}$$

$$= 1 - \frac{\sum_1^n E(X_j^2)}{u^2}. \qquad \textbf{9.4.2}$$

If expression **9.4.2** were a bound simply for $P(|S_n| \leqslant u)$ it would give no more than the Chebichev inequality of section 2.7. What is interesting is that **9.4.2** gives a lower bound for the probability that $S_1, S_2, \ldots, S_n$ are *simultaneously* less than u in modulus. Furthermore, the result is not derived on the assumption of independence, but on the very much weaker assumption **9.4.1**: that the conditional mean should have the same value as the unconditional mean $E(X_j)$, which is zero.

Denote the event

$$[|S_j| < u; j = 1, 2, \ldots, n]$$

by A, and the event

$$[|S_j| \geq u \text{ first for } j = i]$$

by B_i. Then $\sum_1^n B_i = \bar{A}$, so that

$$P(A) = 1 - \sum_1^n P(B_i). \tag{9.4.3}$$

Now

$$E(S_n^2) = \sum_1^n P(B_i)E(S_n^2 \mid B_i) + P(A)E(S_n^2 \mid A). \tag{9.4.4}$$

We can bound the conditional expectations from below by

$$E(S_n^2 \mid A) \geq 0, \tag{9.4.5}$$

$$E(S_n^2 \mid B_i) = E[(S_n - S_i)^2 \mid B_i] + 2E[(S_n - S_i)S_i \mid B_i] + E(S_i^2 \mid B_i)$$
$$\geq u^2. \tag{9.4.6}$$

The last inequality follows from the fact that the first term in the right-hand member of **9.4.6** is positive, the second is zero, since

$$E(S_n - S_i \mid X_1, X_2, \ldots, X_i) = 0$$

in virtue of **9.4.1**, and occurrence of B_i implies that $S_i^2 > u^2$. We thus find from **9.4.4**, **9.4.5** and **9.4.6** that

$$E(S_n^2) \geq u^2 \sum_1^n P(B_i), \tag{9.4.7}$$

and this, together with **9.4.3**, proves the derived inequality **9.4.2**.

The same argument can be adapted to give a strengthened result, which we shall find useful.

Theorem 9.4.2 (*The Hájek–Rényi generalization of Kolmogorov's inequality*)

Under the hypothesis **9.4.1**,

$$P(|S_j| \leq u_j, j = 1, 2, \ldots, n) \geq 1 - \sum_1^n \frac{E(X_j^2)}{u_j^2} \tag{9.4.8}$$

for any positive increasing sequence $\{u_j\}$.

We shall refer to this as the KHR inequality: it enables us to consider a variable bound for the S_j.

Consider a form

$$T = \sum_1^n (\alpha_j - \alpha_{j+1}) S_j^2, \qquad\qquad 9.4.9$$

where the α_j constitute a non-negative decreasing sequence, with $\alpha_{n+1} = 0$. Then

$$E(T) = \sum_1^n \alpha_j E(X_j^2)$$

$$\geqslant \sum_1^n P(B_i) E(T \mid B_i), \qquad\qquad 9.4.10$$

where B_i is now the event

$$\left[|S_j| \geqslant u_j \text{ first for } j = i \right]$$

and

$$A = \overline{\sum B_i} = \left[|S_j| < u_j, j = 1, 2, \ldots, n \right].$$

Now, by the same argument as led to 9.4.7 and 9.4.8,

$$E(S_j^2 \mid B_i) \geqslant \begin{cases} 0 & (j < i), \\ u_i^2 & (j \geqslant i), \end{cases} \qquad\qquad 9.4.11$$

so that, from 9.4.10 and 9.4.11

$$\sum_1^n \alpha_j E(X_j^2) \geqslant \sum_{i=1}^n P(B_i) \sum_{j=i}^n (\alpha_j - \alpha_{j+1}) u_i^2$$

$$= \sum_1^n \alpha_i u_i^2 P(B_i). \qquad\qquad 9.4.12$$

Choosing $\alpha_i = \mu_i^{-2}$, we see that the KHR inequality 9.4.8 follows from 9.4.3 and 9.4.12.

These results depend upon the existence of the second moments of the X_j. However, the same argument can be employed under weaker conditions.

Theorem 9.4.3

Let ϕ be a positive symmetric convex function. Then, under the assumptions of Theorem 9.4.2,

$$P(A) = P(|S_j| < u_j; j = 1, 2, \ldots, n)$$

$$\geq 1 - \sum_1^n \frac{E[\phi(S_j)] - E[\phi(S_{j-1})]}{\phi(u_j)}, \qquad \textbf{9.4.13}$$

where we set $\phi(S_0) = 0$. In particular

$$P(A) \geq 1 - 2^{2-r} \sum_1^n \frac{E(|X_j|^r)}{|u_j|^r} \qquad \textbf{9.4.14}$$

for any r in [1, 2].

We leave the proof to the reader. As before, one considers a form

$$T = \sum (\alpha_j - \alpha_{j+1}) \phi(S_j) \quad \text{with} \quad \alpha_j = \phi(u_j)^{-1}$$

and exploits the fact that, by Jensen's inequality (exercise 2.7.9),

$$E[\phi(S_j) | X_1, X_2, \ldots, X_i] \geq \phi(S_i) \geq \phi(u_i) \quad (j \geq i),$$

if the values of $X_1, X_2, \ldots, X_i$ imply the occurrence of B_i. The second assertion follows by appeal to **6.2.8**.

Exercises

9.4.1   Complete the proof of Theorem 9.4.3.

9.4.2   (Dufresnoy)   Suppose that in addition to **9.4.1** one has the property

$$E(X_j^2 | X_1, X_2, \ldots, X_{j-1}) = E(X_j^2),$$

which is still a weaker assumption than independence. Then the lower inequality in **9.4.11** can be strengthened to

$$E(S_j^2 | B_i) \geq u_i^2 + \sum_{i+1}^j E(X_k^2).$$

Show that an appropriate choice of the α_j in **9.4.9** then yields the following improvement of **9.4.8**:

$$P(A) \geq \prod_{j=1}^n \left(1 - \frac{E(X_j^2)}{u_j^2}\right)$$

provided

$$u_j^2 \geq u_{j-1}^2 + E(X_j^2) \quad (j = 1, 2, \ldots), \quad u_0 = 0.$$

9.5   The laws of large numbers

We have from the beginning emphasized the importance of the idea that a sample average

$$\bar{X}_n = \frac{1}{n} \sum_1^n X_j \qquad \textbf{9.5.1}$$

will, at least under commonly fulfilled conditions, converge in some sense to the common expectation value $E(X_j) = \mu$. In a certain sense this is not the most interesting type of limit, since the limit value is a constant rather than itself a non-degenerate random variable. On the other hand, this is a central type of result, because it was the empirical convergence of mean values that provided a physical starting-point for our theory. We shall continue to assume variables standardized so that $E(X_j) = \mu = 0$.

It has already been established in Theorem 6.2.1 that if condition **9.4.1** holds, and

$$n^{-r} \sum_{1}^{n} E(|X_j|^r) \to 0 \qquad\qquad \textbf{9.5.2}$$

as n increases (for some r in $[1, 2]$) then

$$X_n \xrightarrow{L_r} 0,$$

so that, in particular,

$$X_n \xrightarrow{P} 0. \qquad\qquad \textbf{9.5.3}$$

Relation **9.5.2** would certainly hold if, for example, $E(|X_j|^r)$ were bounded uniformly in j for some $r > 1$.

The assertion **9.5.3** is known as the *weak law of large numbers*. It can be proved under even more general conditions, but we have already obtained, by elementary means, quite a satisfactory degree of generality. It is interesting that we have not needed to assume the X_j independently or identically distributed : conditions **9.4.1** and **9.5.2** are sufficient.

However, it turns out that an appeal to the inequalities of the last section yields the stronger result $X_n \xrightarrow{\text{a.c.}} 0$ under only slightly stronger conditions.

Theorem 9.5.1 (*The strong law of large numbers*)

Suppose that the summands X_j obey the condition **9.4.1** *and that*

$$\sum_{1}^{\infty} \frac{E(|X_j|^r)}{j^r} < \infty \qquad\qquad \textbf{9.5.4}$$

for some r in $[1, 2]$. Then $X_n \xrightarrow{\text{a.c.}} 0$. In particular, the strong law holds if the X_j obey **9.4.1** *and if $E(|X_j|^r)$ is bounded uniformly in j for some r greater than unity.*

The actual necessary and sufficient condition for independent identically distributed X_j is $E(|X|) < \infty$, so that we can obtain a result quite near the best in this case by relatively simple means. Condition **9.5.4** implies **9.5.2** (see exercise 9.5.1) but is in fact only slightly stronger.

To prove the result, consider partial sums

$$T_j = S_{n+j-1} = S_n + X_{n+1} + X_{n+2} + \ldots + X_{n+j-1} \quad (j = 1, 2, \ldots),$$

so that the event

$$C_{nm} = \left[|\bar{X}_j| < \varepsilon; j = n, n+1, \ldots, n+m\right]$$

is identical with the event

$$\left[|T_j| < (j+n-1)\varepsilon; j = 1, 2, \ldots, m\right].$$

Now apply Theorem 9.4.3 to the sequence of partial sums T_j, the summands being regarded as $S_n, X_{n+1}, X_{n+2}, \ldots$. (If the reader feels that the proof of Theorem 9.4.3 is too sketchy, he can restrict himself to the case $r = 2$, which is proved in full in Theorem 9.4.2). We obtain, from **9.4.14**,

$$P(C_{nm}) \geqslant 1 - \frac{2^{2-r}}{(n\varepsilon)^r} \sum_1^n \beta_j - 2^{2-r} \sum_{n+1}^m \frac{\beta_j}{(j\varepsilon)^r},$$

where we have written $E(|X_j|^r)$ as β_j for convenience. Letting m tend to infinity, we have then

$$P(|\bar{X}_j| < \varepsilon; j \geqslant n) \geqslant 1 - \frac{4}{(2\varepsilon)^r}\left[\frac{1}{n^r}\sum_1^n \beta_j + \sum_{n+1}^\infty \frac{\beta_j}{j^r}\right]. \qquad 9.5.5$$

As shown in exercise 9.5.1, the final bracket in **9.5.5** tends to zero with increasing n if **9.5.4** holds. The convergence **9.5.4** is thus a sufficient condition (together with **9.4.1**) for the validity of the strong law.

If $E(|X_j|^r)$ has a uniform (in j) bound for some $r > 1$, then exercise 2.7.10 shows that this also holds for some r in the range $(1,2]$. The convergence **9.5.4** is then obvious.

Exercises

9.5.1   Note that

$$\frac{1}{n^r}\sum_1^n \beta_j \leqslant \left(\frac{k}{n}\right)^r \sum_1^k \frac{\beta_j}{j^r} + \sum_{k+1}^n \frac{\beta_j}{j^r},$$

if $1 \leqslant k \leqslant n$. Show by an appropriate choice of k that this expression converges to zero with increasing n if **9.5.4** holds.

9.5.2   Show that, if the summands X_j obey condition **9.4.1** and have a uniformly bounded absolute rth moment, r having a value in $(1, 2]$, then

$$n^\gamma \bar{X}_n \xrightarrow{\text{a.c.}} 0,$$

where γ is any index less than $1 - 1/r$.

9.6 Martingale convergence, and applications

Let us make a change of notation, and set

$$Y_n = \sum_1^n X_j. \qquad \textbf{9.6.1}$$

Then the condition **9.4.1**, which has proved so consistently important, can be written

$$E(Y_{n+1} \mid Y_1, Y_2, \ldots, Y_n) = Y_n \quad (n = 1, 2, \ldots). \qquad \textbf{9.6.2}$$

A sequence $\{Y_n\}$ obeying the condition **9.6.2** is termed a *martingale*. It is not a very obvious object for study, but it does in fact turn out to be ubiquitous and important. The following theorem indicates part of the reason.

Theorem 9.6.1

Let $\{\mathscr{B}_n\}$ *be an increasing sequence of fields, and let*

$$Y_n = E^{\mathscr{B}_n}(Z) \qquad \textbf{9.6.3}$$

for a fixed r.v. Z. Then $\{Y_n\}$ *is a martingale, and*

$$E(Y_n^2) \leqslant E(Z^2). \qquad \textbf{9.6.4}$$

The converse result, that a martingale has the representation **9.6.3**, can be proved under mild conditions. We have used the notation $E^{\mathscr{B}}(Z)$ introduced at the end of section 5.4. The sequence $\{\mathscr{B}_n\}$ is increasing if members of $\mathscr{B}_n$ are also members of $\mathscr{B}_{n+1}$.

By writing $E^{\mathscr{B}_n}(Z)$ as $E^{\mathscr{B}_n}[E^{\mathscr{B}_{n+1}}(Z)]$ we see that

$$Y_n = E^{\mathscr{B}_n}(Y_{n+1}). \qquad \textbf{9.6.5}$$

Since Y_n is a r.v. defined on $\mathscr{B}_n$, specification of the values of $Y_1, Y_2, \ldots, Y_n$ is an event in $\mathscr{B}_n$. Equation **9.6.2** thus follows as a special case of **9.6.5**, so that **9.6.3** does indeed define a martingale.

Now, recall the interpretation of a conditional expectation as a least square approximant (Theorem 5.3.1): $Y_n = E^{\mathscr{B}_n}(Z)$ is the r.v. Z_n defined on $\mathscr{B}_n$ which minimizes $E[(Z_n - Z)^2]$. It then follows (exercise 5.7.4) that Y_n is orthogonal to $Z - Y_n$, so that

$$E(Z^2) = E(Y_n^2) + E[(Z - Y_n)^2], \qquad \textbf{9.6.o}$$

whence **9.6.4** follows. Of course, statement **9.6.4** has content only if $E(Z^2)$ is finite.

One can thus regard a martingale (or, at least, a large class of them) as a sequence of least square approximants to a r.v. Z, based on an ever-increasing field of functions. This gives the concept some constructive meaning.

Theorem 9.6.2

If $\{Y_n\}$ is a martingale for which $E(Y_n^2)$ is uniformly bounded, then it converges almost certainly.

This is obviously a key result, and is quickly proved. The condition on $E(Y_n^2)$ is a natural one, in view of **9.6.4** (although this can be relaxed: see exercise 9.6.1). The theorem is also interesting because $\{Y_n\}$ has a limit Y which is in general a genuine r.v. and not just a constant, as in Theorem 9.5.1. In the case **9.6.3** the limit Y must be interpretable as $E^{\mathscr{B}\infty}(Z)$: this would be Z itself if Z belonged to $\mathscr{B}_\infty$.

We find from the martingale property **9.6.2** that

$$E(Y_n^2) = \sum_1^n E(X_j^2),$$

9.6.7

so the condition on $E(Y_n^2)$ implies that

$$\sum_1^\infty E(X_j^2) < \infty.$$

9.6.8

Now, it follows from Kolmogorov's inequality **9.4.2** that

$$P(|Y_m - Y_n| < \varepsilon; m > n) = P\left(\left|\sum_{n+1}^m X_j\right| < \varepsilon; m > n\right)$$

$$\geq 1 - \varepsilon^{-2} \sum_{n+1}^\infty E(X_j^2).$$

9.6.9

Relation **9.6.8** implies that this lower bound tends to unity as $n \to \infty$, so that $\{Y_n\}$ is a.c. mutually convergent, and consequently a.c. convergent.

For a situation which. rather unexpectedly, provides an example of a martingale, consider the branching process $\{X_n\}$ of section 6.6, in the case where the population tends to increase, so that $\alpha > 1$. (Note that the X_n bear no relation to the summands in **9.6.1**: the conflict in notation is unfortunate, but short-lived.) Now consider the normalized population size, on the assumption that the population size is initially of size K:

$$Y_n = \frac{X_n}{E(X_n)} = \frac{X_n}{K\alpha^n}.$$

9.6.10

It follows from **6.6.10** and **6.6.14** that

$$E(Y_{n+1} \mid Y_1, Y_2, \ldots, Y_n) = E(Y_{n+1} \mid Y_n) = Y_n,$$

9.6.11

$$E(Y_n^2) \leq \frac{\beta}{K\alpha(\alpha - 1)},$$

9.6.12

as we leave the reader to verify. Thus $\{Y_n\}$ is a martingale, of bounded mean square if β is finite, and so is a.c. convergent.

The intuitive reason for such convergence is that the population size varies erratically while it is small, but the growth rate stabilizes as the population becomes larger. So the variation of Y_n stems principally from the variation of the population while it is small, and has not yet stabilized its growth rate, either by becoming very large, or by becoming extinct. Graphs of sample sequences demonstrate this behaviour very clearly (see Harris, 1963, p. 12).

Another example of a martingale is in the field of *statistical estimation*. Suppose we can take observations $\xi_1, \xi_2, \ldots$, which we assume to be independent and $N(\mu, 1)$, and that we wish to estimate the unknown mean μ from these observations. Suppose that $\hat{\mu}_n$ is the estimate based upon $\xi_1, \xi_2, \ldots, \xi_n$; we shall choose it as the function of these variables that minimizes $E[(\hat{\mu}_n - \mu)^2]$. In order to obtain a sensible extremal problem we shall assume that μ is itself a r.v. say $N(\alpha, \beta^{-1})$, and that E averages over both ξ- and μ-variations. (The introduction of this *prior distribution* can be justified, in some cases at least.)

By its construction $\{\hat{\mu}_n\}$ is a martingale, and so is a.c. convergent to a limit $\hat{\mu}$ if $E(\mu^2) < \infty$, or $\beta > 0$. The question is, whether $\hat{\mu}$ can be identified with μ. We leave it to the reader to verify that

$$\hat{\mu}_n = \frac{\alpha\beta + \sum_1^n \xi_j}{\beta + n} \qquad\qquad 9.6.13$$

and $\quad E[(\hat{\mu}_n - \mu)^2] = \dfrac{1}{\beta + n} \qquad\qquad 9.6.14$

so that $\hat{\mu}_n \xrightarrow{\text{m.s.}} \mu$. The a.c. limit $\hat{\mu}$ and the m.s. limit μ can be identified, with probability one, since

$$|\hat{\mu} - \mu| \leqslant |\hat{\mu}_n - \hat{\mu}| + |\hat{\mu}_n - \mu|.$$

A final example is found in the treatment of the concept of conditional expectation itself. If we consider a sequence $E^{\mathscr{B}_n}(Z)$ conditioned by an increasing sequence of finite fields $\mathscr{B}_n$, then these conditional expectations are well defined by the elementary, constructive approach of section 5.1. However, because they constitute a martingale they tend to an a.c. limit $E^{\mathscr{B}_\infty}(Z)$ even if $\mathscr{B}_\infty$ is infinite. We thus have a convergent construction for a conditional expectation on an infinite field. One should now verify that the limit Y is essentially independent of the particular sequence $\{\mathscr{B}_n\}$ taken to reach $\mathscr{B}_\infty$, and that it satisfies the equation equivalent to **5.4.3**:

$$E(\zeta Z) = E(\zeta Y) \cdot \qquad\qquad 9.6.15$$

for any r.v. ζ defined on $\mathscr{B}_\infty$. We shall not pursue these points, but they are not difficult to prove by exploitation of the fact that $\{E^{\mathscr{B}_n}(Z)\}$ is a sequence of least square approximants (cf. exercise 5.4.2). It is interesting that we can establish constructively the existence of a solution to **9.6.15** without the usual appeal to the notion of a Radon–Nikodym derivative.

9.6.1 Demonstrate Theorem 9.6.2 for the case where $E(|Y_n|^r)$ is uniformly bounded for some $r \geqslant 1$.

9.6.2 The Y_n of the branching process is so strongly convergent that we can use much cruder methods. Show that

$$\sum \gamma^{2n} E(|Y_{n+1} - Y_n|^2) < \infty$$

and hence that

$$\gamma^n (Y_n - Y) \xrightarrow{\text{a.c.}} 0 \quad \text{if} \quad |\gamma| < \alpha^{\frac{1}{2}},$$

without appeal to the martingale property.

9.6.3 Show that the c.f. $\phi(\theta)$ of Y in the branching process example **9.6.10** must obey the equation

$$\phi(\alpha\theta) = G[\phi(\theta)].$$

and that $\phi'(1) = 1$.

9.6.4 Consider the situation of Theorem 9.6.1 and suppose that Y_n has the a.c. limit Y. Show that **9.6.6** extends to

$$E(Z^2) = E(Y_n^2) + E[(Y - Y_n)^2] + E[(Z - Y)^2]$$

and hence that $Y_n \xrightarrow{\text{m.s.}} Y$.

9.6.5 Consider the problem of estimating θ, analogous to that of estimating μ by **9.6.13**, from $\xi_1, \xi_2, \ldots, \xi_n$. Here the ξ_j are assumed to be independently distributed with density

$$\theta^{-1} e^{-\xi/\theta} \quad (\xi \geqslant 0)$$

and θ has prior density

$$\theta^{-m} e^{-\alpha\theta^{-1}} \quad (\theta > 0).$$

9.7 Convergence in rth mean

The notion of L_r-convergence is important in practice (the case $r = 2$ being a particularly natural one to work with) and important for the *extension* problem of the next two chapters (where the case $r = 1$ presents itself first).

First, a few definitions. We shall say that X belongs to L_r if $E(|X|^r) < \infty$. Thus the extent of L_r depends upon the particular process (i.e. the sample space and probability distribution upon it) that one is considering. It is often convenient to work with the *norm* of X:

$$\|X\| = [E(|X|^r)]^{1/r}. \qquad \textbf{9.7.1}$$

This depends upon r, but if we work with a fixed r there is rarely any need to indicate the dependence.

189 Convergence in rth mean

Two r.v.s X and X' for which

$$E(|X - X'|^r) = 0 \qquad \text{9.7.2}$$

are said to be L_r-*equivalent*.

Two useful inequalities are the c_r-*inequality*

$$|X + Y|^r \leqslant c_r[|X|^r + |Y|^r], \qquad \text{9.7.3}$$

where

$$c_r = \begin{cases} 1 & (0 \leqslant r \leqslant 1), \\ 2^{r-1} & (r \geqslant 1), \end{cases} \qquad \text{9.7.4}$$

and the *Minkowski* inequality

$$\left\| \sum_1^n X_j \right\| \leqslant \sum_1^n \left\| X_j \right\| \qquad (r \geqslant 1) \qquad \text{9.7.5}$$

(see exercise 9.7.1).

Theorem 9.7.1

If $X_n \xrightarrow{L_r} X$, *then* X *belongs to* L_r *and*

$$E(|X|^r) = \lim E(|X_n|^r).$$

Because, if $r \leqslant 1$, then by the c_r-inequality

$$|E(|X_n|^r) - E(|X|^r)| \leqslant E(|X_n - X|^r) \to 0 \qquad \text{9.7.6}$$

and, if $r > 1$, we see from the Minkowski inequality that

$$|(\|X_n\| - \|X\|)| \leqslant \|X_n - X\| \to 0. \qquad \text{9.7.7}$$

However, the key theorem is the following:

Theorem 9.7.2

$\{X_n\}$ *is* L_r-*convergent if and only if it is* L_r *mutually convergent.*

The second assertion follows directly from the c_r-inequality

$$E(|X_m - X_n|^r) \leqslant c_r E(|X_m - X|^r) + c_r E(|X_n - X|^r). \qquad \text{9.7.8}$$

It is the direct assertion which is rather more tricky to prove, but extremely useful: that mutual convergence implies convergence.

Consider first the case $r = 1$, which is of especial interest for the next chapter. The treatment is readily generalized to the case $r \geqslant 1$, with which we shall content ourselves.

Let us assume that

$$\sum_1^\infty \varepsilon_n < \infty, \qquad\qquad\qquad\qquad \textbf{9.7.9}$$

where $\varepsilon_n = E(|X_{n+1} - X_n|)$. $\qquad\qquad\qquad\qquad$ **9.7.10**

This can always be achieved by considering a subsequence of $\{X_n\}$, if necessary.
    Consider now the r.v.s

$$\underline{X_{nm}} = X_n - \sum_n^{m-1} |X_{j+1} - X_j|,$$

$$\overline{X_{nm}} = X_n + \sum_n^{m-1} |X_{j+1} - X_j|, \qquad\qquad \textbf{9.7.11}$$

for which $\underline{X_{nm}} \leqslant X_j \leqslant \overline{X_{nm}} \quad (n \leqslant j \leqslant m)$, $\qquad$ **9.7.12**

and consider also

$$\underline{X_n} = \lim_{m \to \infty} \underline{X_{nm}}$$

$$\overline{X_n} = \lim_{m \to \infty} \overline{X_{nm}}. \qquad\qquad\qquad\qquad \textbf{9.7.13}$$

Since the variables **9.7.11** are monotone in m,

$$E(\underline{X_n}) = \lim_m E(\underline{X_{nm}})$$

by Axiom 5 of section 2.2, and similarly for $E(\overline{X_n})$, so that for any r.v. X satisfying

$$\underline{X_n} \leqslant X \leqslant \overline{X_n} \qquad\qquad\qquad\qquad \textbf{9.7.14}$$

we have

$$E(X_n) - \sum_n^\infty \varepsilon_j \leqslant E(X) \leqslant E(X_n) + \sum_n^\infty \varepsilon_j. \qquad \textbf{9.7.15}$$

In particular, **9.7.15** is satisfied by $X = X_j$ for $j \geqslant n$.
    Now the class of r.v.s X satisfying **9.7.14** for *all* n is not empty, since $\underline{X_n} \leqslant \overline{X_n}$, and the two sequences are respectively monotone increasing and monotone decreasing. Because $E(X)$ will then satisfy **9.7.15** for all n, we see by **9.7.9** that

$$E(X) = \lim E(X_n). \qquad\qquad\qquad\qquad \textbf{9.7.16}$$

Furthermore,

$$E(|X - X_n|) \leqslant E(\overline{X_n} - \underline{X_n}) = 2\sum_n^\infty \varepsilon_j \to 0 \qquad \textbf{9.7.17}$$

so that X is an L_1-limit of $\{X_n\}$, which exists by construction. Note also that if X and X' are any two r.v.s satisfying **9.7.14** for all n then

$$E(|X - X'|) \leqslant E(\overline{X_n} - \underline{X_n}) \to 0, \qquad\qquad \textbf{9.7.18}$$

so that these two L_1-limits are L_1-equivalent.

To complete the proof, we have merely to take account of the fact that, in order to fulfil condition **9.7.9**, we may have established L_1-convergence only for a subsequence. If X_n is any member of the original sequence, and $X_{n'}$ the nearest member of the subsequence, then

$$E(|X_n - X|) \leqslant E(|X_n - X_{n'}|) + E(|X_{n'} - X|), \qquad\qquad 9.7.19$$

and both these quantities tend to zero with increasing n. The r.v. X is then an L_1 limit of the original sequence.

We adapt the proof to the case $r \geqslant 1$ by appealing to the Minskowski inequality **9.7.5**. We now define

$$\varepsilon_n = \|X_{n+1} - X_n\| \qquad\qquad 9.7.20$$

and demand **9.7.9**, as before. By suitable application of the Minkowski inequality we then find that, for any X and X' satisfying **9.7.14**,

$$\|X_n\| - \sum_n^\infty \varepsilon_j \leqslant \|X\| \leqslant \|X_n\| + \sum_n^\infty \varepsilon_j \qquad\qquad 9.7.21$$

and $\quad \|X - X'\| \leqslant 2 \sum_n^\infty \varepsilon_j. \qquad\qquad 9.7.22$

The details of the proof are easily completed.

Theorem 9.7.2 gives us an immediate criterion for convergence in a number of useful cases. We give two examples.

Theorem 9.7.3

Let the r.v.s $\xi_1, \xi_2, \ldots$ have zero mean and be uncorrelated. The sum $\sum_1^\infty \xi_j$ is L_2-convergent iff $\sum E(\xi_j^2) < \infty$.

For, if $X_n = \sum_1^n \xi_j$, then

$$E[(X_m - X_n)^2] = \sum_{n+1}^m E(\xi_j^2) \quad (m \geqslant n) \qquad\qquad 9.7.23$$

and the necessary and sufficient condition for this to tend to zero as m and $n \to \infty$ is that the sum $\sum E(\xi_j^2)$ should converge. So, for example, the sum **9.1.1** is L_2-convergent if and only if $|\rho| < 1$.

Theorem 9.7.4

Let X_n be the least square approximant (either linear or unrestricted) to a r.v. Z based on a family of r.v.s $\mathscr{F}_n$. If $\{\mathscr{F}_n\}$ is increasing and $E(Z^2) < \infty$ then $\{X_n\}$ is L_2-convergent.

This is the analogue of Theorem 9.6.2, stated for L_2-convergence rather than a.c. convergence, and applicable also to linear least square approximating sequences. If $\xi_n = X_n - X_{n-1}$, then the ξ_n are uncorrelated, and

$$\sum E(\xi_n^2) \leqslant E(Z^2).$$

Application of Theorem 9.7.3 then proves the result.

Exercise

9.7.1   *Proof of Minkowski's inequality*   Show by direct minimization of the first member with respect to X that

$$|X|^r - r\theta|XY| \geqslant (1-r)|\theta Y|^s,$$

where $r > 1$, and $r^{-1} + s^{-1} = 1$. Taking expectations over X and Y, and minimizing the upper bound for $E(|XY|)$ thus obtained with respect to θ, show that

$$E(|XY|) \leqslant [E(|X|^r)]^{1/r}[E(|Y|^s)]^{1/s}.$$

(This is Hölder's inequality, a generalization of the Cauchy inequality.) Show also that equality is attainable.

Alternatively this result can be written as

$$[E(|X|^r)]^{1/r} = \max_Y E(|XY|),$$

where the r.v. Y is restricted by $E(|Y|^s) = 1$. Show that Minkowski's inequality follows from this identity and the relation

$$\max_Y E(|\sum X_j Y|) \leqslant \sum_j \max_Y E(|X_j Y|).$$

10 Extension

10.1 Extension of the expectation functional: the finite case

We return now to the problem enunciated in section 2.1 : given the expectations of a number of r.v.s $\{X_\nu\}$, what bounds can be placed on $E(Y)$ for another r.v. Y? This might be regarded as the *extension* problem in its wider sense; the problem in its narrower sense is to delimit the class of r.v.s Y for which $E(Y)$ is exactly determined by the given expectations. The second sense is the more usual, and if we refer to the 'extension problem' without qualification, this will be the one we mean. Note that this extension is not a matter of 'convention'; that is, we do not merely assign a value to $E(Y)$ which is consistent with previously known expectation values, and attractive on some other grounds. The value is *determined* by the axioms, and there is no latitude for choice.

Almost everything we have done could be characterized as the expression of some expectations in terms of others, so the extension idea is fundamental. We shall develop the basic extension theorems in this chapter, and apply them in the next.

Let us assume to begin with that we know the expectation values of a finite number of r.v.s $X_0, X_1, \ldots, X_m$, where $X_0 = 1$. We shall assume that these values have been assigned in accord with Axioms 1–5 of section 2.2, and shall not go into the question of testing for consistency, although this is obviously important.

We know then the expectation of any linear combination

$$X = c_0 + \sum_1^m c_j X_j. \qquad \qquad 10.1.1$$

Let the field of r.v.s denoted in this way be denoted by $\mathscr{F}$. Now, if X belongs to $\mathscr{F}$, and if $Y \leqslant X$ (i.e. $Y(\omega) \leqslant X(\omega)$ for $\omega \in \Omega$), then certainly $E(Y) \leqslant E(X)$. It is in just this way that we obtained the Chebichev inequalities of section 2.7. Now we can attempt to improve this upper bound for $E(Y)$ by choosing the X of $\mathscr{F}$ for which $E(X)$ is least, while still requiring that $X \geqslant Y$. In fact this procedure gives us the sharp upper bound for $E(Y)$.

Theorem 10.1.1

(i) *The sharp upper and lower bounds for $E(Y)$ are respectively*

$$\bar{E}(Y) = \min_{\substack{X \in \mathscr{F} \\ X \geqslant Y}} E(X), \qquad\qquad\qquad\qquad \textbf{10.1.2}$$

$$\underline{E}(Y) = \max_{\substack{X \in \mathscr{F} \\ X \leqslant Y}} E(X).$$

(ii) *The extremal distribution for which* **10.1.2** *is attained is one concentrated on the values ω of Ω for which $X(\omega) = Y(\omega)$, X being the r.v. giving the extreme in* **10.1.2**.

(iii) *This distribution can be concentrated on at most $m+1$ points of Ω.*

(iv) *$E(Y)$ is completely determined just for those r.v.s for which $\underline{X} \leqslant Y \leqslant \bar{X}$, where $\underline{X}$ and $\bar{X}$ belong to $\mathscr{F}$, and $E(\underline{X}) = E(\bar{X})$.*

The first assertion is the key one. Expression **10.1.2** is obviously an upper bound; the point of the theorem is that it is a sharp bound, i.e. there exists a distribution on Ω consistent with the given expectations for which this bound can be attained.

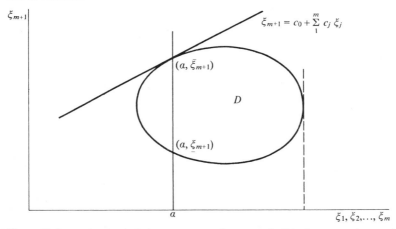

Figure 7 Determination of the extreme values permissible for an unknown expectation

We shall set $Y = X_{m+1}$ for notational convenience. Consider now the space S of Theorem 3.3.1, but in $m+1$ dimensions (see Figure 7). As before, the convex region D is the set of points representable as

$$\xi_j = E(X_j) \quad (j = 1, 2, \ldots, m+1),$$

where E is an expectation corresponding to some probability distribution on Ω. Let the given values of $E(X_j)$ be denoted by $\alpha_j (j = 1, 2, \ldots, m)$. Then, if these

195 **Extension of the expectation functional: the finite case**

values have been consistently assigned, the line $\xi_j = \alpha_j \, (j = 1, 2, \ldots, m)$ will intersect D, and the extreme ξ_{m+1} values, $\underline{\xi}_{m+1}$ and $\overline{\xi}_{m+1}$, in which this line intersects D will be the least and greatest attainable values of $E(Y)$.

At the point $(\alpha_1, \alpha_2, \ldots, \alpha_m, \overline{\xi}_{m+1})$ on the boundary of D there will be a supporting hyperplane, say

$$\xi_{m+1} = c_0 + \sum_1^m c_j \xi_j, \qquad\qquad\qquad \textbf{10.1.4}$$

so that $\quad \overline{\xi}_{m+1} = c_0 + \sum_1^m c_j \alpha_j = c_0 + \sum_1^m c_j E(X_j) = E(X),$ $\qquad\qquad$ **10.1.5**

say. Since D lies on one side of the hyperplane **10.1.4**, the same side as $(\alpha_1, \alpha_2, \ldots, \alpha_m, \xi_{m+1})$ for $\xi_{m+1} < \overline{\xi}_{m+1}$, we see that all points in D satisfy

$$\xi_{m+1} \leqslant c_0 + \sum_1^m c_j \xi_j. \qquad\qquad\qquad \textbf{10.1.6}$$

Choosing the values of ξ corresponding to a one-point distribution on Ω, we find that

$$Y(\omega) \leqslant c_0 + \sum_1^m c_j X_j(\omega) = X(\omega) \quad (\omega \in \Omega), \qquad\qquad \textbf{10.1.7}$$

since the points $\xi_j = X_j(\omega) \, (j = 1, 2, \ldots, m + 1)$ (corresponding to the one-point distributions) certainly lie in D. But **10.1.5** and **10.1.7** imply that for some distribution consistent with the given expectations it is possible to find an X in $\mathscr{F}$ for which $X \geqslant Y$ and $E(Y) = E(X)$. The bound **10.1.2** is thus sharp, and **10.1.3** similarly.

Assertion (i) solves the broader extension problem in principle, and its immediate corollary (iv) similarly solves the narrower extension problem.

Assertion (ii) follows from **10.1.7**: if we average this inequality over ω we can obtain equality of the two resulting expectations only if the average is restricted to ω-values for which there is equality.

Assertion (iii) follows from the fact that any point on the boundary of an $(m+1)$-dimensional convex region is an average of at most $m+1$ extreme points of the region, and the extreme points in this case are those derived from the one-point distributions.

In writing the equation of the hyperplane in the form **10.1.4** we have tacitly assumed the coefficient of ξ_{m+1} to be non-zero, i.e. that the hyperplane is not parallel to the ξ_{m+1}-axis. The contrary case is usually that in which $\underline{\xi}_{m+1} = \overline{\xi}_{m+1}$ (see the dotted line in the diagram) and $E(Y)$ is fully determined. The simplest course is to consider the situation as a limit of the previous one, by considering a sequence of α-values which lead one to this extreme case.

There are virtually no general methods for finding the X which gives the sharp bound in special cases. If by ingenuity one has found an $X \geqslant Y$ which one suspects yields the sharp bound, one can best establish its sharpness by finding a distribution consistent with the given expectations for which $E(X) = E(Y)$. For example, we have noted that the Markov and Chebichev inequalities **2.7.2** and **2.7.6** are sharp, if these bounds are less than unity.

Exercises

10.1.1   Suppose the r.v. X is known to lie in the interval $(0, b)$, and to have expectation μ. Show that

$$P(X > a) \geqslant \max\left(0, \frac{\mu - a}{b - a}\right),$$

and that this inequality is sharp.

10.1.2   Suppose that the values of $E(X)$ and $E(X^2)$ are known. Show that

$$P(X > a) \leqslant \frac{\text{var}(X)}{\text{var}(X) + [a - E(X)]^2},$$

for $a \geqslant E(X)$, and that the inequality is sharp in this range.

10.1.3   Consider the profit function $g_N(X)$ defined by the first formula of section 2.5. Show that if the full distribution of demand X is not known, but only $E(X) = \mu$ and $\text{var}(X) = \sigma^2$, then the sharp lower bound for $G_N = E[g_N(X)]$ is

$$G_N \geqslant a\mu + \tfrac{1}{2}(a - b + c)(N - \mu) - \tfrac{1}{2}(a + b + c)\sqrt{[(N - \mu)^2 + \sigma^2]}$$

or $$G_N \geqslant -c\mu + \frac{(a + c)\mu^2 - b\sigma^2}{\mu^2 + \sigma^2} N$$

according as $\sigma^2 + \mu^2$ is less than or greater than $2N\mu$.

If one had to optimize stocks on the basis of this limited information, it would be reasonable to choose the value of N which maximizes this lower bound. Show that this is

$$N = \begin{cases} \mu + \dfrac{(a - b + c)\sigma}{2\sqrt{[b(a + c)]}} & \left(\dfrac{\sigma}{\mu} < \sqrt{\dfrac{a + c}{b}}\right), \\ 0 & \left(\dfrac{\sigma}{\mu} > \sqrt{\dfrac{a + c}{b}}\right), \end{cases}$$

with corresponding values $a\mu - \sigma\sqrt{[b(a + c)]}$ and $-c\mu$ for the bound.

The discontinuous behaviour of N is interesting, when compared with the case of full information.

10.2   Generalities on the infinite case

In general the set $\{X_\nu\}$ will be infinite. Let us append to this all r.v.s whose expectations are *immediately* determined from the quantities $E(X_\nu)$ by

Axioms 1–5 of section 2.2, so that we have a field of r.v.s which is closed under finite linear operations, and under monotone limits. Other extensions may be possible, along the lines of the previous section, but this we have yet to determine.

It is the possibility of considering limits that introduces a new feature. Let us note that, if the prescribed expectation values are consistent with Axioms 1–4, then Axiom 5 is self-consistent. In other words, if one has two monotone sequences tending to the same limit, then the two corresponding sequences of expectations will also tend to the same limit.

For, suppose that the two monotone non-decreasing sequences $\{X_n\}$ and $\{X'_n\}$ have a common limit X. Then, for fixed m and variable n, $X_n - X'_m$ is a non-decreasing sequence with a non-negative limit, so that by Axioms 1 and 5
$$\lim_n E(X_n - X'_m) \geqslant 0$$

or $\lim E(X_n) \geqslant E(X'_m).$ **10.2.1**

(We exclude the case $E(X'_m) = +\infty$, when it is plain that both sequences of expectations have $+\infty$ as limit). Taking the limit of large m in **10.2.1**, we see that

$$\lim E(X_n) \geqslant \lim E(X'_n).$$

The same proof justifies the reverse inequality, so that the two limits are equal, and can be unequivocally identified with $E(X)$. It is because we have restricted ourselves to monotone limits that such identification is possible, but one would like to be able to draw similar conclusions for more general limit sequences.

We give the assertions corresponding to (i) and (iv) of Theorem 10.1.1, which now hold in a slightly weakened form.

Theorem 10.2.1

(i) *The sharp upper and lower bounds for $E(Y)$ are respectively*

$$\bar{E}(Y) = \inf_{\substack{X \in \mathscr{F} \\ X \geqslant Y}} E(X),$$ **10.2.2**

$$\underline{E}(Y) = \sup_{\substack{X \in \mathscr{F} \\ X \leqslant Y}} E(X).$$ **10.2.3**

(ii) $E(Y)$ *is completely determined just for those r.v.s for which sequences $\{\underline{X}_n\}$ and $\{\overline{X}_n\}$ in $\mathscr{F}$ exist such that*

$$\underline{X}_n \leqslant Y \leqslant \overline{X}_n \quad and \quad E(\overline{X}_n - \underline{X}_n) \to 0.$$

The r.v. Y is thus an L_1 limit of either sequence.

The point of the notation sup and inf rather than max and min is that the extreme expectations **10.2.2** and **10.2.3** may not actually be attained for any X in $\mathscr{F}$, although one can find X-sequences for which the infimum or supremum is approached arbitrarily closely.

The supporting hyperplane proof goes through as before, although lack of closure of D means that one can only make the weaker statement of Theorem 10.2.1. Assertion (ii) is a similarly weakened version of Theorem 10.1.1(iv). To prove the final statement, we observe that

$$E|\overline{X_n} - Y| \leqslant E(\overline{X_n - X_m})$$

for all m, and that the right-hand member tends to zero as m and $n \to \infty$. Note that Y is an L_1-limit of $\{\overline{X_n}\}$ (and of $\{\underline{X_n}\}$) for *all* distributions on Ω consistent with the given expectations.

One can scarcely progress further without making explicit assumptions concerning the family of basic random variables $\mathscr{F}$. We shall discuss two particular cases in sections 10.3 and 10.4.

10.3   **Extension on a linear lattice**

We know that $\mathscr{F}$ is closed under finite linear operations, and under monotone limits. Let us suppose that it is also closed under the taking of moduli, so that $|X|$ belongs to $\mathscr{F}$ if X does. Such a family of functions is termed a *linear lattice*.
The additional assumption implies that if X_1 and X_2 belong to $\mathscr{F}$, then so do

$$X_1 \vee X_2 = \max(X_1, X_2) \quad \text{and} \quad X_1 \wedge X_2 = \min(X_1, X_2).$$

This follows from the formulae

$$X_1 \vee X_2 = \tfrac{1}{2}(X_1 + X_2 + |X_1 - X_2|),$$
$$X_1 \wedge X_2 = \tfrac{1}{2}(X_1 + X_2 - |X_1 - X_2|).$$

10.3.1

The assumption that $|X|$ belongs to $\mathscr{F}$ if X does is certainly convenient, and in certain cases reasonable. Its main use is that it provides us with a distance function

$$\|X - Y\| = E(|X - Y|),$$

10.3.2

with which we can measure the effective deviation between two elements of $\mathscr{F}$: this in turn enables us to be explicit about the limit behaviour of more general sequences than the monotone sequences.
As an example, suppose two indicator functions I_A and I_B belong to $\mathscr{F}$, so that we know the values of $P(A)$ and $P(B)$. Then, since $I_A \wedge I_B = I_{AB}$, we also know the value of $P(AB)$. The additional assumption is thus closely related to the 'intersection assumption' of section 3.5; that if A and B belong to the relevant field of events, then so does AB.
In fact, the case discussed in section 3.5 is just that which gives us our principal application of the results of this section (see exercise 10.3.1 and section 11.1). Suppose that the sample space is the real line, representing the

values of a r.v. X. Suppose also that we know the distribution function $F(x)$ of X; i.e. we know the probability measure of the sets $\{X \leqslant x\}$ for variable x. It is then true that, given $P(A)$ and $P(B)$ we can calculate $P(AB)$, and the sets $\{X \leqslant x\}$ generate a field. The extension problem is then: for what functions H can one evaluate $E[H(X)]$, on the basis of knowledge of $F(x)$?

Let us now return to the general problem. The treatment becomes simpler if we reduce $\mathscr{F}$ to $\mathscr{F}'$, the set of members X of $\mathscr{F}$ for which $E(|X|) < \infty$. What we shall now prove is that the maximal extension of $\mathscr{F}'$ is obtained by appending to $\mathscr{F}'$ the L_1-limits of L_1 Cauchy sequences in $\mathscr{F}'$; i.e. of sequences $\{X_n\}$ for which

$$E(|X_m - X_n|) \to 0. \qquad \qquad \textbf{10.3.3}$$

Theorem 10.3.1

Any L_1 Cauchy sequence in $\mathscr{F}'$, $\{X_n\}$, is L_1-convergent to a class of L_1-equivalent r.v.s X for which $E(X) = \lim E(X_n)$. By appending such limits to $\mathscr{F}'$ we obtain a consistent extension of $\mathscr{F}'$ which is also maximal.

All these assertions have already been proved in one way or another. We know from Theorem 9.7.2 that an L_1 Cauchy sequence has an L_1-limit, with expectation $\lim E(X_n)$. As this theorem emphasizes, and as is obvious from the constructive proof of Theorem 9.7.2, the 'limit' is really a class of r.v.s, all L_1-equivalent. One point that can be made is that the r.v.s $\underline{X}_n$ and $\overline{X}_n$ defined in **9.7.13** themselves belong to $\mathscr{F}'$, so that in introducing them we remain on known ground. Terms such as 'L_1-convergent' and 'L_1-equivalent' are valid for *all* distributions on Ω consistent with the given expectations.

The extension is maximal, since we know from Theorem 10.2.1 that any element in the maximal extension is an L_1-limit of a sequence in $\mathscr{F}'$. (More than that, it is the L_1-limit of sequences approaching it from either side, as were the limits in the proof of Theorem 9.7.2.) Being maximal, it is then itself closed under all the operations we have used: linear combination, the taking of moduli, and the taking of L_1-limits.

To prove consistency, we must demonstrate facts such as that the sum of the L_1-limits of $\{X_n\}$ and $\{Y_n\}$ is the L_1-limit of $\{X_n + Y_{n'}\}$, or that if two sequences $\{X_n\}$ and $\{X_n'\}$ both have X as L_1-limit, then $\lim E(X_n) = \lim E(X_n')$. This is straightforward, and we leave verification to the reader.

Theorem 10.3.2

Theorem 10.3.1 still holds if one does not know the r.v.s X as functions $X(\omega)$ of ω, but simply the expectation values $E(X)$ $(X \in \mathscr{F}')$.

This is surprising, because in the deduction of the basic extension Theorems 10.1.1 and 10.2.1 it was essential that we should know all r.v.s as explicit functions on Ω. However, the fact that a linear lattice is such a rich class means that we can 'relate' two r.v.s X and Y of Ω (see exercise 10.3.2) and so can do just as well as if we knew them as functions of ω. On the other hand, we must know that a representation of a r.v. as a function on Ω exists in principle, or the convergence proof of Theorem 9.7.2 would be invalid.

The theorem is immediate: since we can evaluate $E(|X_m - X_n|)$, we can still construct or recognize Cauchy sequences. We can thus achieve the same extension as before, and, as this is now based upon less information, it is *a fortiori* maximal.

However, the two cases differ when it comes to identification of limits. If the r.v.s are known as functions of ω, then the limit of an L_1 Cauchy sequence $\{X_n\}$ in $\mathscr{F}'$ can always be identified with a r.v. (or r.v.s) X, by the construction of Theorem 9.7.2. However, for the case of Theorem 10.3.2 one cannot always make such an identification. The limit of an L_1 Cauchy sequence in $\mathscr{F}'$ is always meaningful, but can be identified with a specific r.v. only if the limit also lies in $\mathscr{F}'$.

Exercises

10.3.1 Suppose the linear lattice $\mathscr{F}$ is generated from a r.v. X and the constant r.v. 1. Show that $(X - x)_+$ (defined in **2.2.4**) belongs to $\mathscr{F}$. Show also, by consideration of the function

$$J(X) = \lim_{K \to \infty} \min[1, K(X - x)_+]$$

that the indicator function of the set $X \leqslant x$ belongs to $\mathscr{F}$, so that we can evaluate the distribution function of F.

10.3.2 Suppose $\mathscr{F}$ is generated from X, Y and 1. Show heuristically that we can determine a probability such as

$$P[a < H(X, Y) \leqslant b],$$

and so detect a possible functional relationship between X and Y.

10.4 Extension on a quadratic field

Another natural situation is the following. Suppose that for all r.v.s X of a class $\mathscr{G}$ we know the *expected products*

$$E(X_\mu X_\nu) \quad (X_\mu, X_\nu \in \mathscr{G}).$$

We shall not assume that the r.v.s are known explicitly as functions on Ω (although it is assumed that they can be so represented in principle): the only way different r.v.s can be related is by using what information the mean products can yield.

This is just about the minimal information that will enable one to detect any relationship at all between r.v.s, and so the case is interesting for just that reason. The given expectations allow us to apply the methods of linear least square approximation (sections 2.6 and 5.7) and the natural distance function is

$$\|X - Y\| = \sqrt{E[(X - Y)^2]}.$$
10.4.1

From the given expectations we can calculate mean products of finite linear combinations of elements of $\mathscr{G}$, and so obtain an immediate extension. We shall assume this extension performed, so that $\mathscr{G}$ includes all finite linear combinations of its elements. However, we cannot use the operation of taking monotone limits, because the elements are not known as functions on Ω, so that a sequence cannot be verified to be monotone. As in the previous section, it is convenient to restrict $\mathscr{G}$ to $\mathscr{G}'$, the class of elements X of $\mathscr{G}$ for which $E(X^2) < \infty$.

We obtain a further extension of $\mathscr{G}'$ (maximal, as it will turn out) by appending to $\mathscr{G}'$ the L_2-limits of sequences in $\mathscr{G}$. (We should recall that the terms 'L_2-convergence' and 'm.s. convergence' are synonymous).

Theorem 10.4.1

If $\{X_n\}$ *is an* L_2 *Cauchy sequence in* $\mathscr{G}'$, *i.e.*

$$E[(X_m - X_n)^2] \to 0,$$
10.4.2

then it is L_2-*convergent to a limit* X, *and*

$$E(X^2) = \lim E(X_n^2) \qquad E(XZ) = \lim E(X_n Z),$$

for Z *in* $\mathscr{G}'$. *The extension* $\mathscr{G}''$ *of* $\mathscr{G}'$ *obtained by appending these limits is closed under the formation of* L_2-*limits.*

The first statement, and that concerning $E(X^2)$, follow from Theorem 9.7.2. The statement concerning $E(XZ)$ follows from the Cauchy inequality

$$|E[(X - X_n)Z]|^2 \leq E(Z^2)E[(X - X_n)^2] \to 0.$$
10.4.3

To see that $\mathscr{G}''$ is closed, note that if X is the L_2-limit of a sequence $\{X_n\}$ in $\mathscr{G}''$, then X_n must be representable as the L_2-limit of a sequence $\{X_{n1}, X_{n2}, \ldots\}$ in $\mathscr{G}'$. We leave the reader to verify that X is also the L_2-limit of the $\mathscr{G}'$ sequence $\{X_{nn}\}$, and hence belongs itself to $\mathscr{G}''$.

Theorem 10.4.2

The extension $\mathscr{G}''$ *of Theorem* 10.4.1 *is maximal in that, if* $E(Y^2)$ *is determined by knowledge of* $E(YZ)$ $(Z \in \mathscr{G}')$, *then* Y *must belong to* $\mathscr{G}''$.

This is the analogue of Theorem 10.2.1 in that it characterizes the maximal extension possible under the present assumptions.

We have to assume knowledge of the expectations $E(YZ)$ in order to be able to relate Y to the elements of $\mathscr{G}'$ at all. Find the element X of $\mathscr{G}''$ which minimizes $E[(Y - X)^2]$. Then

$$Y = X + (Y - X),$$
10.4.4

where $Y - X$ is orthogonal to all elements of $\mathscr{G}''$.

Now define

$$Y^* = X + \alpha(Y - X),\qquad\qquad\qquad \textbf{10.4.5}$$

where α is an arbitrary constant. Then

$$E(Y^*Z) = E(YZ)$$

for Z in $\mathscr{G}'$ (in fact, in $\mathscr{G}''$), but

$$E[(Y^*)^2] = E(X^2) + \alpha^2\, E[(Y - X)^2].\qquad\qquad\qquad \textbf{10.4.6}$$

Thus, if $E[(Y - X)^2] > 0$, we have found a r.v. Y^* possessing the same mean products with the elements of $\mathscr{G}'$ as Y does, but whose mean square is indeterminate. If the mean square is not to be indeterminate, we must have $E[(Y - X)^2] = 0$, so that Y belongs to $\mathscr{G}''$.

11 Examples of Extension

11.1 **Integrable functions of a scalar random variable**

Consider the problem touched upon in sections 3.5 and 10.3: given the distribution function $F(x)$ of a r.v. X, for what functions H can we deduce the value of $E[H(X)]$? In particular, for what sets A can we deduce $P(X \in A)$? Such functions H are termed *integrable*, and such sets A are said to be *measurable*. This is obviously an extension problem, starting from the given expectations $F(x)$. Furthermore, as noted in section 10.3, it can be regarded as an extension problem on the linear lattice generated from the initial pair of elements 1 and X.

The application of countably many operations of linear combination and the taking of moduli to the indicator functions $I_{X \leqslant x}$ will generate the class of *simple functions*; that is, those functions for which the X-axis can be divided into a countable number of intervals, on each of which the function is constant. We know then from Theorem 10.3.1 that $H(X)$ is integrable if and only if we can deduce from the axioms that

$$E(|H - H_n|) \to 0, \tag{11.1.1}$$

where $\{H_n(X)\}$ is a sequence of simple functions.

Let us restrict our attention to positive functions, allowing the possibility that $E(X) = +\infty$. The general case then follows, provided we avoid the case where we are led to the indeterminate evaluation $+\infty - \infty$.

Theorem 11.1.1

$E[H(X)]$ *is certainly determined by* $F(x)$ *and equal to* $\int H(x)\,dF(x)$ *if* H *is a continuous function, or a monotone limit of simple functions* (*a Borel function*).

Consider first the case of simple H: suppose $H(X)$ takes the value h_j on the interval $(a_{j-1} < X \leqslant a_j)$, these intervals constituting a decomposition of the real axis. It follows then from the axioms that, if I_j is the indicator function of this interval, then

$$E(H) = \sum_j h_j E(I_j)$$

$$= \sum_j h_j [F(a_j) - F(a_{j-1})]. \tag{11.1.2}$$

Turning now to continuous H, let A_j be the X-set satisfying

$$\frac{j-1}{n} < H(X) \leqslant \frac{j}{n},$$
11.1.3

where n is a fixed positive integer. Then A_j is a countable set of intervals, so that we can evaluate $P(X \in A_j)$. If $H_n(X)$ is the simple function taking the value j/n on A_j, we have then

$$H_n(X) - \frac{1}{n} \leqslant H(X) \leqslant H_n(X)$$

so that $\quad E(H_n) - \frac{1}{n} \leqslant E(H) \leqslant E(H_n),$ **11.1.4**

where $E(H_n)$ is evaluated by

$$E(H_n) = \sum_j \frac{j}{n} P(X \in A_j).$$
11.1.5

The expectation of $E(H)$ is thus evaluated to within $1/n$ by **11.1.4** and **11.1.5**. As we let n increase, $E(H_n)$ will decrease and $E(H_n) - 1/n$ will increase, and they will converge to a common value which must be the value of $E(H)$. Obviously

$$E(|H - H_n|) \leqslant \frac{1}{n}.$$

The final assertion of the theorem is a direct application of Axiom 5 (section 2.2): the main point of the theorem is then that the Borel functions include the continuous functions.

Our initial set $\mathscr{F}$ was essentially the class of simple functions of X; we have extended it to the class $\mathscr{B}$ of Borel functions of X. This is a great deal less than the maximal extension $\mathscr{F}_1$ which we know to be possible: a fact demonstrated by the observation that the maximal extension $\mathscr{F}_1$ depends upon the given expectations (i.e. upon the distribution of X), while the extension to $\mathscr{B}$ is one that is possible for *all* $F(x)$. Of course, it is useful to have a 'universal' extension of this kind, and the Borel functions are adequate for most purposes, but one should keep in mind that a much wider extension is possible: to the class of all functions H which are L_1-limits of simple functions.

While on this subject, we should note that $\mathscr{F}_1$ (as an extension of $\mathscr{F}'$) must in general be *smaller* than the class of r.v.s with finite absolute expectation on a *given fixed* distribution. One can see two reasons for this, which are in fact equivalent. For one thing, we started from a restricted set of r.v.s $\mathscr{F}'$, and so cannot in general generate all functions of L_1 from it. For another, the r.v.s of $\mathscr{F}_1$ must have finite absolute expectation for *all* distributions on Ω consistent with the given expectations.

The convergence theorems of classical measure and integration theory are rather different in character to those of probability theory, because they emphasize more the study of $X(\omega)$ itself as a function, than the study of its 'statistical' behaviour under varying averaging operations. A classic problem is this: if $\{X_n\}$ is a sequence of integrable functions of ω converging pointwise to a limit X, under what conditions is X integrable and $E(X) = \lim E(X_n)$? One sufficient condition (by axiom) is that $\{X_n\}$ be monotone. Another is that $Y \leqslant X_n \leqslant Z$ where Y and Z are integrable (or, as a special case, $|X_n| \leqslant Y$ with integrable Y). This is the *dominated convergence theorem* of which we have given a short probabilistic proof in Theorem 9.3.3. That proof would not be acceptable in the present context, because we cannot *a priori* regard $E(|X - X_n|)$ and $P(|X - X_n| > \varepsilon)$ as well-defined quantities. The treatment can be made satisfactory; however, we indicate the conventional direct proof in exercises 11.1.3 and 4.

Exercises

11.1.1   Suppose X positive. Show that $E(X)$ is finite if and only if $\int x \, dF(x)$ is, and that then the two are equal. (Use equations **11.1.2** and **11.1.3**.)

11.1.2   Consider a sequence of r.v.s $\{X_n\}$, and define

$$\overline{X}_n = \sup_{m \geqslant n} X_m,$$

$$\underline{X}_n = \inf_{m \geqslant n} X_m.$$

Show that if $\{X_n\}$ converges pointwise (for each given ω) to a *finite* limit X, then $\{\overline{X}_n\}$ and $\{\underline{X}_n\}$ converge monotonely to X (from above and below respectively).

11.1.3   *Fatou's lemma*   One defines $\liminf X_n$ and $\limsup X_n$ as $\lim \underline{X}_n$ and $\lim \overline{X}_n$ respectively. Show from $\overline{X}_n \geqslant \underline{X}_n$ that

$$\liminf E(X_n) \geqslant E(\liminf X_n).$$

11.1.4   *The dominated convergence theorem*   Prove this theorem (see text) by applying Fatou's lemma to the sequences $\{X_n - Y\}$ and $\{Z - X_n\}$.

11.2   Expectations derivable from the characteristic function

So many of our results have been expressed in terms of c.f.s that it is natural to ask: for what functions $H(X)$ is $E[H(X)]$ determinable from knowledge of $\phi(\theta) = E(e^{i\theta X})$ for real θ? This is an extension problem, and related to it is the query that arose in section 8.5: if $\phi_n(\theta)$ is the c.f. of X_n, and $\phi_n(\theta) \to \phi(\theta)$, for what functions H can one assert that

$$E[H(X_n)] \to E[H(X)]?$$

The structure of $\phi(\theta)$ makes it natural to regard this problem as one of extension on a quadratic field (section 10.4), since if

$$\xi(\theta) = e^{i\theta X} \qquad\qquad\qquad \textbf{11.2.1}$$

then we are given the expectations

$$E[\xi(\theta)\overline{\xi(\theta')}] = \phi(\theta - \theta'). \qquad\qquad\qquad \textbf{11.2.2}$$

Our basic field $\mathscr{G}$ consists thus of the trigonometric sums $\sum a_j e^{i\theta_j X}$, and we know from theorem 10.4.1 that we can evaluate $E[H(X)]$, and indeed $E[H(X)e^{-i\theta X}]$ and $E(|H|^2)$ for any function H for which we can show that

$$E(|H - H_n|^2) \to 0, \qquad\qquad\qquad \textbf{11.2.3}$$

where $\{H_n(X)\}$ is a sequence of trigonometric sums. The fact that the r.v.s $\xi(\theta)$ are complex is no real complication: for complex r.v.s the theory of section 10.4 is still valid if we assume knowledge of $E(X_\mu \overline{X_\nu})$ rather than of $E(X_\mu X_\nu)$ $(X_\mu, X_\nu \in \mathscr{G})$, and define $\|X\| = [E(|X|)^2]^{\frac{1}{2}}$.

However, in taking this mean square approach we are neglecting one important piece of information: that the $\xi(\theta)$ are known as functions of X, by **11.2.1**. This point does not invalidate Theorem 10.4.1. However, it does mean that we should be able to supplement the argument of that theorem, and obtain stronger results: for example, that $E(H)$ is the limit of a sequence $E(H_n)$, although in a weaker sense than that of **11.2.3**.

As an example of the pure mean square approach, we give the following.

Theorem 11.2.1

If $\phi''(0)$ exists, then

$$E(X^2) = -\phi''(0), \qquad\qquad\qquad \textbf{11.2.4}$$

$$E(Xe^{iX\theta}) = -i\phi'(\theta), \qquad\qquad\qquad \textbf{11.2.5}$$

and all these quantities are finite.

We know from Theorem 8.4.3 that if $E(X^2)$ is finite then $-\phi''(0)$ exists and is equal to it, so the theorem provides a partial converse to this result. The proof is outlined in exercises 11.2.1 and 2.

We shall now turn our attention to more general results which exploit the functional form of $\xi(\theta)$.

Theorem 11.2.2

Suppose that $H(x)$ has the Fourier representation

$$H(x) = \int h(\theta)e^{i\theta x}\, d\theta, \qquad\qquad\qquad \textbf{11.2.6}$$

then $\quad E[H(X)] = \int h(\theta)\phi(\theta)\, d\theta \qquad\qquad\qquad \textbf{11.2.7}$

provided $h(\theta)$ is absolutely integrable.

Equation **11.2.7** follows formally from **11.2.6** if we take expectations under the integral sign. This commutation of expectation and integral will be valid if

$$E\left[\int |h(\theta)e^{i\theta X}| \, d\theta\right] < \infty$$

(Fubini's theorem; see Kingman and Taylor, 1966, p. 144), which it will be if

$$\int |h(\theta)| \, d\theta < \infty.$$

However, this theorem puts conditions on H which are excessively strong, and not very explicit. One would like simple and fairly undemanding conditions directly on H itself, which would ensure the validity of some version of **11.2.7**. The finding of these inevitably involves one in Fourier theory to a certain extent.

We can note a few helpful points. Firstly, if H is absolutely integrable, that is, $\int |H(x)| \, dx < \infty$, then

$$h(\theta) = \frac{1}{2\pi} \int e^{-i\theta x} H(x) \, dx \qquad\qquad \textbf{11.2.8}$$

exists, and is bounded by $(1/2\pi)\int |H(x)| \, dx$. Secondly, $h(\theta)$ will converge to zero as $\theta \to \pm\infty$ at a rate determined by the degree of continuity of H. For example, if H and its derivatives up to order p exist and are absolutely integrable, then repeated partial integration of **11.2.8** shows that $h(\theta) = O(\theta^{-p})$ for large θ. For $p = 2$, this would ensure absolute integrability of h. Thirdly, if both H and h are absolutely integrable, and H is continuous, then the inversion of **11.2.8** to **11.2.6** is valid (Goldberg, 1961, p. 16).

Thus, if we can 'tailor' a given H so that it and its first two derivatives are absolutely integrable, without changing its expectation by more than a pre-assigned amount, then we can certainly calculate its expectation.

Theorem 11.2.3

Suppose X is finite with probability one, H is continuous, and $E[|H(X)|] < \infty$. Then $E[H(x)]$ is determinable in terms of the c.f. of X.

The proof will give a construction for this determination. Set

$$H_{MN}(X) = \begin{cases} H(X) & (|H(X)| < M, |X| < N), \\ 0 & \text{(otherwise)}. \end{cases} \qquad\qquad \textbf{11.2.9}$$

Because of the first and third assumptions, $E[H_{MN}(X)]$ can be made to differ arbitrarily little from $E[H(X)]$ if M and N are chosen large enough. However, $H_{MN}(X)$ is absolutely integrable, for given finite M and N.

Modify this again to

$$\hat{H}(X) = \int H_{MN}(X + \alpha u) f(u) \, du, \qquad\qquad \textbf{11.2.10}$$

where f is a standard normal density. If H is continuous, then H_{MN} is uniformly continuous, and $\hat{H} - H_{MN}$ can be made arbitrarily and uniformly small if α is chosen small enough. However, for α positive, $\hat{H}$ possesses derivatives of all orders, each of these also being absolutely integrable.

We thus see from the discussion before the theorem that $\hat{H}$ obeys the conditions of Theorem 11.2.2 and that its expectation is equal to $\int \hat{h}(\theta)\phi(\theta)\,d\theta$, in an obvious notation. However, by choosing M and N large enough, and α small enough, we can make $|E(H) - E(\hat{H})|$ arbitrarily small. The expectation $E(H)$ is thus determinable from knowledge of $\phi(\theta)$.

The theorem can be further refined, but is general enough for most purposes as it stands. There is an obvious version of it dealing with the convergence problem of section 8.5.

Theorem 11.2.4

Consider a sequence of r.v.s $\{X_n\}$, whose characteristic functions converge to that of a r.v. X:

$$E(e^{i\theta X_n}) \to E(e^{i\theta X}) \qquad\qquad\qquad \textbf{11.2.11}$$

for any fixed real θ. Then

$$E[H(X_n)] \to E[H(X)] \qquad\qquad\qquad \textbf{11.2.12}$$

provided H is continuous, $P(|X_n| > N)$ converges uniformly (in n) to zero as $N \to \infty$, and $E[|H(X_n)|I_{Mn}]$ converges uniformly (in n) to zero as $M \to \infty$. Here I_{Mn} is the indicator function of the set $|H(X_n)| > M$.

The last two conditions, on the $\{X_n\}$ sequence, are considered also to apply to X. These conditions can be weakened: in fact, the condition on $P(|X_n| > N)$ can be replaced simply by the demand that X be finite with probability one, and this in turn by the requirement that $\phi(\theta) = E(e^{i\theta X})$ equal unity and be continuous at the origin (see exercises 11.2.4 and 6). We thus have

Theorem 11.2.5

*The convergence **11.2.12** is a consequence of **11.2.11** if H is bounded and continuous, $\phi(\theta)$ is continuous and equal to unity at the origin. Moreover, $E(H)$ is then determinable from ϕ.*

Exercises

11.2.1   Consider the sequence

$$H_n(X) = \frac{i}{n}(e^{iX/n} - 1).$$

Show that this is m.s. mutually convergent, and so m.s. convergent, iff $\phi''(0)$ exists.

11.2.2   (Continuation of exercise 11.2.:.) Use the inequality

$$\frac{\sin \theta}{\theta} > \frac{2}{\pi} \quad \text{for} \quad |\theta| < \tfrac{1}{2}\pi$$

to show that $E(X^2)$ exists and equals $\lim E(H_n^2)$, and hence that the m.s. limit of $\{H_n\}$ is m.s. equivalent to its pointwise limit X, and can be identified with it. Theorem 11.2.1 then follows from this identification, and appeal to Theorem 10.4.1.

11.2.3   Show more generally that if $i^{2j}\phi^{(2j)}(0)$ exists, then $E(X^{2j})$ exists, and is equal to it.

11.2.4   Suppose that the value of the next maximum of $(\sin x)/x$ after that at the origin is $1 - \alpha$. and that the smallest value of x for which $(\sin x)/x = 1 - \alpha$ is β (see Figure 8). Then

$$1 - \frac{\sin x}{x} \geqslant \alpha \mathscr{H}\big[|x| - \beta\big],$$

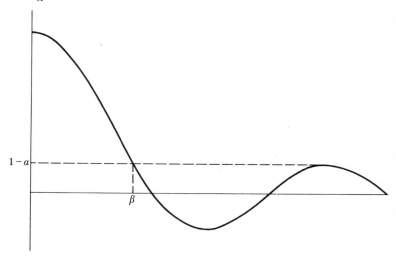

Figure 8 Graph of $(\sin x)/x$

where $\mathscr{H}$ is the Heaviside function of **2.5.1**. Demonstrate from this that, if X is a r.v. with c.f. $\phi(\theta)$, then

$$P(|X| > N) \leqslant \frac{N}{2\alpha\beta} \int_{-\beta N^{-1}}^{\beta N^{-1}} [1 - \phi(\theta)]\, d\theta$$

and hence that X is finite with probability one if ϕ obeys the conditions of Theorem 11.2.5.

11.2.5 As an example of the type of behaviour these conditions on ϕ guard against, consider $X_n = nY$, where Y is a r.v. which may be zero with a certain probability and is otherwise normally distributed, say. Then, for infinite n, X_n is infinite with probability $P(Y \neq 0)$. Show that

$$\phi_n(\theta) \to \begin{cases} 1 & (\theta = 0), \\ P(Y = 0) & (\theta \neq 0). \end{cases}$$

11.2.6 Use the inequality of exercise 11.2.4, and the fact that, by **11.2.11**, $\phi_n \to \phi$ uniformly on any bounded interval, to deduce the weakening of the condition on $P(|X_n| > N)$ indicated after Theorem 11.2.4.

12 Some Interesting Processes

The four sections of this chapter are quite unrelated. They cover four particular processes, each of interest in its own right, and each generating a special set of ideas.

12.1 Quantum mechanics

It is remarkable that one of the most basic and impressive structures of theoretical physics, quantum mechanics, should be intrinsically probabilistic, and yet with a logic deviating essentially from that of the 'classical' probability theory we have studied. The quantum mechanical approach is quite similar to that which we took to classical probability in chapter 2, in its selection of expectation as the basic concept, and in the detailed form of the axioms to which the expectation functional is subject. However, certain differences in the initial formulation produce profound differences in the final structure and conclusions.

We shall have to accept a certain number of basic ideas derived from physical arguments. This once done, we shall find that the theory develops in a self-contained and striking manner.

P1 *To any variable a of physical interest there corresponds a Hermitian linear operator A. (We shall refer to such a variable as an observable, the point being that only quantities which are in principle observable are regarded as having physical interest.)*

P2 *The operator corresponding to a function $f(a)$ of a is $f(A)$.*

P3 *The operator corresponding to a linear function $\sum c_j a_j$ of observables is the same function $\sum c_j A_j$ of the operators, A_j.*

For formal simplicity we shall assume that all the operators A are $n \times n$ matrices, although in actual physical contexts one must consider operators on more general spaces. If the matrix A has spectral representation

$$A = \sum_{j=1}^{n} \alpha_j \phi_j \phi_j^{\dagger}, \qquad\qquad \textbf{12.1.1}$$

then we can define $f(A)$ in a self-consistent fashion by

$$f(A) = \sum f(\alpha_j)\phi_j \phi_j^{\dagger} \qquad\qquad \textbf{12.1.2}$$

provided the quantities $f(\alpha_j)$ are uniquely defined (see exercise 12.1.1).

The expectation value of a is then considered to be a functional $\mathscr{E}$ of the corresponding operator:

$$E(a) = \mathscr{E}(A). \qquad \qquad 12.1.3$$

The functional $\mathscr{E}$ is assumed to have the following properties:

E1 $\mathscr{E}$ is a real scalar.

E2 $\mathscr{E}$ is linear:

$$\mathscr{E}(c_1 A_1 + c_2 A_2) = c_1 \mathscr{E}(A_1) + c_2 \mathscr{E}(A_2), \qquad \qquad 12.1.4$$

where c_1 and c_2 are real coefficients.

E3 $\mathscr{E}(A)$ is non-negative if A is non-negative definite.

E4 $\mathscr{E}(I) = 1,$ $12.1.5$

where I is the identity matrix.

Axioms E1–E4 are obviously very similar to Axioms 1–4 of section 2.2. One can also add a continuity axiom, similar to Axiom 5, but this we shall not need if we consider only the finite structure of the theory. However, the fact that $\mathscr{E}$ has a matrix A as argument, rather than a scalar function $X(\omega)$, makes a great difference. As for Axiom 3 of section 2.2, relation **12.1.4** is regarded as defining $\mathscr{E}(c_1 A_1 + c_2 A_2)$ only if the right-hand member is determinate. Note, in connexion with Axiom E3, that any Hermitian matrix can be written as a difference of non-negative definite matrices, in analogy with the relation $X = X_+ - X_-$ for scalars.

Our first assertion is the following.

Theorem 12.1.1

The expectation functional $\mathscr{E}$ has the form

$$\mathscr{E}(A) = \mathrm{tr}(UA), \qquad \qquad 12.1.6$$

where U is a non-negative definite Hermitian matrix of unit trace.

Here we have used tr( ) to denote the trace of a matrix, that is the sum of its diagonal elements. Statement **12.1.6** corresponds to the assertion in the classical case with finite state space that $E(X)$ must be of the form **2.4.1**, with $\{p_k\}$ a distribution.

Since $\mathscr{E}$ is linear, it can be written explicitly as a linear function of the elements of A; $\sum\sum u_{jk} a_{kj}$. In matrix notation this becomes just representation **12.1.6**. Since A is Hermitian and $\mathscr{E}$ is real, one sees directly that U can be also chosen as Hermitian. The assertion $\mathrm{tr}(U) = 1$ follows from Axiom E4, if one chooses $A = I$ in **12.1.6**. If one chooses $A = \phi\phi^\dagger$ for an arbitrary vector ϕ, then one has, in virtue of Axiom E3,

$$0 \leqslant \mathscr{E}(\phi\phi^\dagger) = \phi^\dagger U \phi, \qquad \qquad 12.1.7$$

so that U is non-negative definite. We have thus established the theorem. The conditions of the theorem are also sufficient; the functional $\mathscr{E}(A)$ constructed there has all the properties E1–E4.

From **12.1.2** and **12.1.6** we have

$$E[f(a)] = \sum_j f(\alpha_j)(\phi_j^\dagger U \phi_j).$$ **12.1.8**

Since f is arbitrary, relation **12.1.8** must have the following interpretation.

Theorem 12.1.2

The observable a can assume only the values α_j, and

$$P(a = \alpha_j) = \phi_j^\dagger U \phi_j,$$ **12.1.9**

where α_j is an eigenvalue of A, and ϕ_j the corresponding normalized ($\phi_j^\dagger \phi_i = 1$) eigenvector ($j = 1, 2, \ldots, n$).

This is one of the most interesting and radical conclusions of quantum theory; it follows, as we see, quite quickly from the axioms. However, stronger statements can be made.

Theorem 12.1.3
Suppose that some observable b is known to have the value β. Then we must have

$$U = \psi\psi^\dagger,$$ **12.1.10**

where ψ is the normalized eigenvector of B corresponding to the eigenvalue β. In the case **12.1.10** *we have then*

$$E(a) = \psi^\dagger A \psi,$$ **12.1.11**

$$and \quad P(a = \alpha_j) = |\phi_j^\dagger \psi|^2.$$ **12.1.12**

This is again one of the basic ideas: that the taking of an observation affects all expectations, which are thereby 'conditioned'. When U has the form **12.1.10** we shall say that the system is in a *pure state*, and that the elements of ψ constitute the *wave function* describing the state. For convenience we shall describe ψ itself as the wave function.

In the general case previously considered, U will have a spectral representation

$$U = \sum_1^n \lambda_j \xi_j \xi_j^\dagger,$$ **12.1.13**

where the λ_j are non-negative and add to unity. We can thus regard this situation as one of *mixed states*, in which the system has wave function ξ_j with probability λ_j ($j = 1, 2, \ldots, n$).

However, in general one works with pure states. The physical justification for this is, largely, that a dynamical system, once in a pure state, and left to itself, will continue in a pure state (see exercise 12.1.3). According to Theorem 12.1.3, a pure state is achieved when the value of some observable has been determined.

To prove Theorem 12.1.3, we note that, since b takes the value β corresponding to eigenvector ψ with probability one, we have, in virtue of **12.1.9**, $\eta^\dagger U \eta = 0$ for every vector η orthogonal to ψ. By **12.1.13** this implies that

$$\sum_j \lambda_j |\xi_j^\dagger \eta|^2 = 0. \qquad\qquad\qquad \textbf{12.1.14}$$

Since all the terms in this sum are non-negative, they must be individually zero. That is, if $\lambda_j > 0$ then $\xi_j^\dagger \eta = 0$ for any vector η orthogonal to ψ, which implies that ξ_j itself must be proportional to ψ. Thus U is of the form constant $\times \psi \psi^\dagger$. Since U has unit trace, and $\psi^\dagger$ is normalized, the constant must be unity. Relations **12.1.11** and **12.1.12** follow from **12.1.6** and **12.1.9** in the case **12.1.10**.

The case **12.1.10** rather corresponds to that of a degenerate distribution in the classical case, when a random variable can take only a single value. However, it is a peculiarity of quantum mechanics that there is no situation corresponding to the idea of a completely degenerate sample space, consisting of a single point, randomness being completely absent.

Theorem 12.1.4

A deterministic situation, in which the values of all observables are known, is impossible if $n > 1$.

In such a case U would certainly have the form **12.1.10**, with ψ an eigenvector of the operator corresponding to an observable of known value. But it is impossible that a given ψ can simultaneously be an eigenvector of *all* Hermitian operators. Indeed, one can obviously make the stronger statement, that one cannot simultaneously know the values of two observables whose operators do not share at least one eigenvector.

Thus some degree of uncertainty is essential in a quantum-mechanical situation – an idea which finds another expression later in the uncertainty relation **12.1.25**.

Such ideas naturally lead one to consider the joint distribution of several observables, and it is notable that quantum mechanics in general gives one no means of doing so, because there is no general rule, corresponding to P2, for forming the operator corresponding to a function $f(a, b)$ of several observables. This would seem to agree with the operational point of view, that if one cannot observe an event (the simultaneous determination of the values of several observables) then there is no point in talking about its probability.

One exception is the case in which operators A and B share the complete set of eigenvectors, so that, if A has spectral representation **12.1.1**, then that of B is of the form

$$B = \sum_j \beta_j \phi_j \phi_j^\dagger. \qquad\qquad\qquad \textbf{12.1.15}$$

In this case the obvious operator to associate with $f(a, b)$ is

$$f(A, B) = \sum_j f(\alpha_j, \beta_j)\phi_j\phi_j^\dagger.$$ **12.1.16**

Corresponding to the interpretation of **12.1.8**, we deduce that in this case the system may be in a pure state, of wave function ϕ_j, with probability $\phi_j^\dagger U \phi_j$, and that a and b are then simultaneously known to have the values α_j and β_j respectively.

In such a case, a and b are said to be *simultaneously observable*. The idea can obviously be generalized to the case of more than two observables.

If A and B do not necessarily share eigenvectors, then **12.1.15** must be replaced by

$$B = \sum_j \beta_j \psi_j \psi_j^\dagger.$$ **12.1.17**

If b is known to have the value β_k then the system will have wave function ψ_k, and by **12.1.12** we have the quantum-mechanical statement of a conditional probability:

$$P(a = \alpha_j | b = \beta_k) = |\phi_j^\dagger \psi_k|^2.$$ **12.1.18**

But the same equation yields

$$P(b = \beta_k | a = \alpha_j) = |\psi_k^\dagger \phi_j|^2 = |\phi_j^\dagger \psi_k|^2,$$

so that we have the following interesting result.

Theorem 12.1.5

Conditional probabilities obey the symmetry relation

$$P(a = \alpha_j | b = \beta_k) = P(b = \beta_k | a = \alpha_j).$$ **12.1.19**

This statement is closely related to the *principle of detailed balance* of statistical mechanics; for further implications see exercise 12.1.6.

Operators A and B will share a full set of eigenvectors if and only if they commute: $AB = BA$. This commutation means that one does not have to distinguish between A^2B, ABA and BA^2, for instance: these are all equal, and all represent the same observable, a^2b. It is in general the commutation, or lack of it, which will tell us whether two variables are or are not simultaneously observable. A rather more explicit result can be obtained as follows: suppose

$$AB - BA = 2iC.$$ **12.1.20**

From the fact that, for real λ,

$$O \leqslant \mathscr{E}[(A + i\lambda B)(A - i\lambda B)]$$ **12.1.21**

$$= \mathscr{E}[A^2 + 2\lambda C + \lambda^2 B^2],$$

we deduce the Cauchy-type inequality

$$E(a^2)E(b^2) \geqslant [\text{tr}(UC)]^2. \hspace{2cm} \textbf{12.1.22}$$

If we replace a by $a - E(a)$, and so A by $A - E(a)I$, and similarly modify b, we do not change the value of the commutator $AB - BA$. Thus **12.1.22** can be refined to

$$\text{var}(a)\,\text{var}(b) \geqslant [\text{tr}(UC)]^2. \hspace{2cm} \textbf{12.1.23}$$

If only C were such that $\text{tr}(UC)$ were independent of U, we would have a universal inequality. This is true of certain pairs of variables a and b called *conjugate observables*, for which the commutator takes the form

$$AB - BA = i\kappa I, \hspace{2cm} \textbf{12.1.24}$$

where κ is a real constant. We have then from **12.1.23**

$$\text{var}(a)\,\text{var}(b) \geqslant \kappa^2. \hspace{2cm} \textbf{12.1.25}$$

This is the so-called *uncertainty principle*, expressing the fact that the two variables a and b can under no circumstances be simultaneously determined.

Actually, a relationship such as **12.1.24** cannot be valid for a finite system such as we have considered, for **12.1.25** would then imply that $\text{var}(b) = \infty$ if a is known, which is impossible in a finite system. However, for an infinite system the conjugacy relation **12.1.24** is possible (see exercise 12.1.7).

Exercises

12.1.1   Deduce from **12.1.1** that **12.1.2** is in fact true if (i) f is a polynomial, (ii) $f(A) = A^{-1}$, (iii) f is a finite power series (with negative as well as positive integral powers) in A.

12.1.2   Confirm that the quantities **12.1.9** are non-negative, and add to unity.

12.1.3   It can be shown from physical arguments that the variation of a system in time is determined by the equation

$$\dot{U} = i\hbar[HU - UH],$$

where $U = U(t)$ is the U of **12.1.6**, now time-dependent, $\dot{U}$ is its rate of change, $\hbar$ is a real constant, and H a particular Hermitian matrix (the *Hamiltonian* operator, corresponding to the observable 'energy'). Suppose that the system is pure at $t = 0$, so that $U(0) = \psi(0)\psi(0)^\dagger$ for a vector $\psi(0)$. Show that it is pure at all times:

$$U(t) = \psi(t)\psi(t)^\dagger,$$

with ψ satisfying

$$\dot{\psi} = i\hbar H\psi.$$

12.1.4   Note that this last equation has the formal solution

$$\psi(t) = e^{i\hbar t H}\psi(0)$$

and the one before it has the formal solution

$$U(t) = e^{ihtH} U(0) e^{-ihtH}.$$

$U(t)$ is thus generated by a similarity transformation based on the unitary matrix $Q(t) = e^{ihtH}$.

12.1.5 Show that the distribution of values of energy does not change with time. Show that the same is true for any observable whose operator commutes with H. (Such an observable is termed an *integral* of the dynamic system.)

12.1.6 Suppose we try formally to construct a joint distribution of values of two observables a and b, consistent with the conditional distributions determined by **12.1.18** and **12.1.19**. Assuming the conditional probabilities **12.1.18** to be non-zero, show that we must necessarily have

$$P(a = \alpha_j) = P(b = \beta_j) = \frac{1}{n} \quad (j = 1, 2, \ldots, n).$$

This corresponds to a mixed state in which $U = n^{-1}I$ in **12.1.6**, and the n possible values (counted as distinct) of *any* observable are equally likely.

12.1.7 Consider the operators A and B on a function $\phi(x)$ defined by

$$A\phi(x) = ih\frac{d\phi}{dx} \qquad B\phi(x) = x\phi(x).$$

If ϕ is square integrable on the real axis, then these are Hermitian, and

$$AB - BA = ihI,$$

so that the two operators (representing *momentum* and *position* respectively) are conjugate.

12.1.8 Show that the inequality **12.1.23** becomes an equality iff $U = \phi\phi^\dagger$, where ϕ satisfies $(A - i\lambda B)\phi = \mu\phi$ for some λ, μ. Hence show that in the preceding example one has equality in **12.1.25** iff $\phi(x)$ has the 'normal' form

$$\phi(x) = \exp(\alpha + \beta x + \gamma x^2),$$

where γ has negative real part. Determine the density functions of position and of momentum in this case.

12.2 **Information theory**

The idea of *amount of information* as a quantitative entity, which can be weighed out like potatoes, was developed principally in communication engineering, and especially by C. E. Shannon (1948). Actually, it is a general concept, which had been in the air for some time, and is closely related to the *entropy* of thermodynamics and statistical mechanics. Nevertheless, it was in communication theory that the new ideas were clearly formulated and developed, although these have since found their way into other areas; notably computing, biology, and the pure mathematics of ergodic theory.

The mathematical quantification of an everyday notion is always a fascinating process; the hopeful opening of a long-concealed door. However, in quantifying, one generally has to narrow down the original concept considerably, and strip it of many of its associations, so hopes and conclusions should be cautious. On the other hand, this reduction is a valuable exercise in itself.

Suppose that one can transmit messages using an alphabet of m letters. These letters may be the dot, dash, and word-space of Morse code ($m = 3$), or the characters available on a teletype machine, or the colours of a set of signal rockets. A message consists of a sequence of such letters, there being a *coding* convention which relates messages and sequences.

One of the basic aims of information theory is to achieve an efficient coding, so that the average lengths of sequences can be reduced to the minimum. In order to be able to average, we must make some assumptions concerning the statistical character of the message.

Let us consider the simplest possible situation, and assume that, in the initial coding, the messages are such that consecutive letters in a message are statistically independent; letter k occurring with probability p_k ($k = 1, 2, \ldots, m$). Then the average information per letter is defined as

$$H = - \sum_{1}^{m} p_k \log p_k. \qquad \text{12.2.1}$$

One can justify this choice of measure of information by arguing from a set of plausible axioms for such a measure. However, we shall appeal rather to the operational justification: definition **12.2.1** suggests itself naturally when we consider the question of efficient coding; the extension of the definition to more general situations is then also plain.

The information H is a positive quantity, which achieves its minimum value of zero when all but one of the p_k are zero, that is when the same letter is always transmitted. At the other extreme, it reaches its maximum value when all the letters are used equally frequently, so that $p_k = 1/m$, and $H = \log m$. This follows from Jensen's inequality (exercise 2.7.9) in the form

$$\frac{1}{m} \sum_{1}^{m} \phi(a_k) \geqslant \phi\left(\frac{1}{m} \sum_{1}^{m} a_k\right). \qquad \text{12.2.2}$$

Setting $a_k = p_k$ and $\phi(x) = x \log x$ we deduce that

$$H = - \sum p_k \log p_k \leqslant -m\left(\frac{1}{m} \sum p_k\right) \log\left(\frac{1}{m} \sum p_k\right) = \log m. \qquad \text{12.2.3}$$

It is plausible that a symbol should carry no information when its nature can be perfectly predicted beforehand, and that it should carry maximum information when the uncertainty of prediction is greatest.

We shall achieve a more efficient coding (in all but the case of maximal H) by considering a long sequence of n letters, and then recoding the m^n n-letter sequences, using the old m-letter alphabet, in such a way that the more probable sequences are transmitted by short sequences in the new coding. Suppose that in this way the expected length of message is reduced from n to ρn. Then ρ is referred to as the *compression coefficient*.

Theorem 12.2.1

Consider the case of indefinitely large n. Then the compression coefficient has lower bound

$$\rho_{\min} = \frac{H}{H_{\max}} = -\frac{\sum p_k \log p_k}{\log m}$$

12.2.4

and this bound can be approached arbitrarily closely, by appropriate recoding.

It is at this point, then, that the information measure H makes its natural appearance.

Consider a sequence of letters $\{k_1, k_2, \ldots, k_n\}$; this will have probability

$$P(k) = \sum_{j=1}^{n} p_{k_j}.$$

Now, consider P itself as a r.v., or, rather,

$$L = -\log P = \sum_{1}^{n} X_j,$$

12.2.6

where the X_j are independent r.v.s taking the value $-\log p_k$ with probability p_k $(k = 1, 2, \ldots, m)$. We have then

$$E(L) = nH,$$

12.2.7

with H defined by **12.2.1**.

Now, we can appeal to the law of large numbers (section 9.5) and deduce that, for large enough n,

$$P\left[\left|\frac{L}{n} - H\right| \leqslant \varepsilon\right] \geqslant 1 - \delta,$$

12.2.8

for arbitrarily small positive ε and δ. Let the sequences for which $|L/n - H| \leqslant \varepsilon$ be termed the ε-*standard sequences*. Then, by **12.2.8**, the probability that a sequence is ε-standard can be made greater than $1 - \delta$. Now the number of ε-standard sequences must lie between $(1 - \delta)/Q_1$ and $1/Q_2$, where Q_1 and Q_2 are upper and lower bounds on the probability of any individual ε-standard sequence. But for an ε-standard sequence

$$e^{-n(H + \varepsilon)} \leqslant P \leqslant e^{-n(H - \varepsilon)},$$

so that the number of ε-standard sequences lies between

$$(1 - \delta)e^{n(H - \varepsilon)} \quad \text{and} \quad e^{n(H + \varepsilon)}.$$

If n' would be the length of message needed to encode just the ε-standard sequences, then

$$(1-\delta)e^{n(H-\varepsilon)} \leqslant m^{n'} \leqslant e^{n(H+\varepsilon)}. \qquad \textbf{12.2.9}$$

If the remaining sequences are coded by n-letter sequences, we have then achieved a compression coefficient of

$$\frac{(1-\delta)n'+\delta n}{n} \leqslant \frac{(1-\delta)(H+\varepsilon)}{\log m} + \delta. \qquad \textbf{12.2.10}$$

Seeing that ε and δ may be chosen arbitrarily small, this coefficient can be made smaller than $\rho_{\min}+\eta$, for arbitrarily small positive η, where $\rho_{\min}$ is given by **12.2.4**.

We have thus proved the second part of the theorem. The first assertion, that $\rho_{\min}$ represents the smallest asymptotic compression coefficient attainable, follows from the facts that

$$\frac{(1-\delta)n'+\delta n}{n} \geqslant \frac{(1-\delta)}{\log m}\left[H-\varepsilon+\frac{\log(1-\delta)}{n}\right]+\delta, \qquad \textbf{12.2.11}$$

and that we have used a recoding which approximates the most efficient one. We leave the reader to fill in the details here.

We have not distinguished between the numbers of sequences of *exactly n* letters and of *at most n* letters; these would be respectively m^n and

$$\sum_{0}^{n} m^r = \frac{m^{n+1}-1}{m-1} < \left(\frac{m}{m-1}\right)m^n.$$

If we recoded m^n equally likely sequences by using sequences of length 1, 2, ... up to n' then we would have

$$\left(\frac{m}{m-1}\right)m^{n'} > m^n > m^{n'},$$

$$\text{or} \quad 1 > \frac{n'}{n} > 1+\frac{1}{n}\log_m\left(1-\frac{1}{m}\right),$$

and so we would achieve no additional compression, asymptotically. The distinction is thus not significant, at the level of approximation to which we are working.

The assertion of the theorem, that the least possible asymptotic compression coefficient is $H/H_{\max}$, is one that holds under very much more general conditions. The method of proof makes it plain that, if there is an appropriate measure of information, it must be

$$H = \lim_{n \to \infty}\left[-\frac{1}{n}E[\log P(k)]\right]. \qquad \textbf{12.2.12}$$

For instance, suppose that in the original coding the letters follow a Markov process with transition matrix (p_{jk}), and asymptotic occupation probabilities p_j. Then, if there are n_{jk} transitions from j to k in the sequence we have

$$H = -\lim \frac{1}{n} E\left(\sum_j \sum_k n_{jk} \log p_{jk}\right)$$
$$= -\sum_j \sum_k p_j p_{jk} \log p_{jk}. \qquad\qquad \textbf{12.2.13}$$

We would expect expression **12.2.13** to be less than expression **12.2.1** with the p_j identified in the two cases (see exercise 12.2.1), because the presence of serial dependence must help one to predict future letters in the sequence, and make it possible to achieve a greater compression upon recoding. This is a point which often proves confusing: the serial dependence implies that observations already taken (letters already read) carry *more* information about the next observation than in the case of independence; the actual taking of the observation must consequently yield *less* information. As an extreme example, the letter 'q' in the English language is invariably followed by a 'u'; the actual observation of the 'u' must thus be informationless, and to transmit the combination 'qu' as a single character 'q' would certainly shorten messages.

The full theory of information and coding goes on to introduce another very important idea, that of transmission over a noisy channel. That is to say, a message sequence is fed into a transmission channel (e.g. a telephone line, or a human brain) and, because of disturbances *en route*, the sequence received at the other end of the channel may differ from that transmitted. In order to ensure high probability of correct transmission a message must now be coded in a redundant form, that is, with enough 'repetition', in some sense, that the original message can still be inferred almost certainly from the distorted one. In aiming at optimal coding under these circumstances, one arrives at a notion closely related to the information content of a message: the information-carrying *capacity* of a channel. We refer the interested reader to Feinstein (1958).

Exercises

12.2.1 Show from Jensen's inequality and the equilibrium equations for the p_j in the Markov process associated with **12.2.13** that

$$\sum_j p_j p_{jk} \log p_{jk} \geqslant p_k \log p_k,$$

and hence that expression **12.2.13** is not greater than expression **12.2.1**.

12.2.2 Suppose that letters in a message are independent, letter k appearing with probability p_k. Suppose, however, that the recoding procedure of the theorem is carried out on the erroneous assumption that letter k appears with probability p_k' ($k = 1, 2, \ldots, m$). Show that the asymptotic compression coefficient thus achieved is

$$\frac{-\sum p_k \log p_k'}{\log m}.$$

Since this coding cannot be optimal, it follows that

$$\sum p_k \log \frac{p_k}{p_k'} \geq 0. \tag{12.2.14}$$

12.2.3   Demonstrate **12.2.14** directly, and show that equality is possible only if $p_k' = p_k$ $(k = 1, 2, \ldots, m)$. Expression **12.2.14** is often used as a measure of deviation between the two distributions. For example, if X and Y are discretely distributed r.v.s with

$$p_{jk} = P(X = x_j, Y = y_k),$$

then $\displaystyle\sum_j \sum_k p_{jk} \log \frac{p_{jk}}{p_{j\cdot}\, p_{\cdot k}}$

is sometimes used as a measure of statistical dependence between the two variables, since

$$p_{j\cdot}\, p_{\cdot k} = \left(\sum_r p_{jr}\right)\left(\sum_s p_{sk}\right)$$

would describe the joint distribution if X and Y were independent.

12.2.4   The information in a distribution with a density must be infinite: explain why, intuitively. However, an 'information difference' such as

$$\int f(x) \log[f(x)/g(x)]\, dx$$

will be meaningful, where f and g are densities. Show that this is invariant under a non-singular transformation of the variable x.

12.2.5   Suppose a r.v. X with probability density f has prescribed mean and variance. Show that, if f is such that $\int f \log f\, dx$ is maximal, then X is normally distributed.

12.3   Dynamic programming; stock control

Let us return to the newsagent's stock optimization problem of section 2.4. As we pointed out then, the problem would have been quite different had the stock been of such a nature that it could be carried over from day to day, because then one day's transactions and decisions would have affected those of the next. Let us consider such a case, typical of the important and growing area of *sequential decision theory*.

   Suppose we consider the newsagent's sales of cigarettes rather than of newspapers, so that stocks can very well be carried over. For simplicity, we suppose there is a single brand, in packets of standard size. We suppose that the agent is left with a stock (or *inventory*, to use the American term) of y_{t-1} packets on the evening of the $(t-1)$th day, that he then always has the opportunity to order a further N_t from the wholesaler before the beginning of the tth day, and that he sells X_t on the tth day. Thus

$$y_t = y_{t-1} + N_t - X_t. \qquad \textbf{12.3.1}$$

We must now separate the costs a little more carefully than we did in section 2.4. Suppose the cost of buying a batch of N packets from the wholesaler is $bN + d_N$, where

$$d_N = \begin{cases} 0 & (N = 0), \\ d & (N > 0). \end{cases} \qquad \textbf{12.3.2}$$

We are thus imagining that b is the cost price of a packet, and d the cost of having an order delivered, whatever its size.

We shall again suppose that there is profit a on the sale of a packet, so that the selling price is $a + b$. We shall also assume that, if the newsagent is left with a stock y in the evening, then there is an immediate cost $A(y)$, which might, for example, have the form

$$A(y) = \begin{cases} c'y & (y \geqslant 0), \\ -cy & (y \leqslant 0). \end{cases} \qquad \textbf{12.3.3}$$

Here c' is the cost per packet of holding a positive stock, and stems from items such as storage, spoilage, and the immobilization of capital. The cost $c|y|$ associated with negative stock is that produced when demand exceeds initial stock. We shall assume for simplicity that demand is always satisfied, and that negative stock represents a situation in which the newsagent has had to satisfy this demand from a secondary source (e.g. by buying quickly from another retailer) and c is the additional expense per packet of doing so. The situation in which stocks can never be negative, and excess demand must simply remain unsatisfied, is slightly more complicated, but can be treated by the methods of this section (see exercise 12.3.1).

Let $F_n(y)$ be the expected total future profit the newsagent will make under an optimal ordering policy, from a point in time when he has a stock of y, and intends to continue operations for a further n days. We shall assume that $F_0(y) = 0$, i.e. that once operations cease all books are cleared, and no further income or expenses are expected. We shall assume initially that the demands X_t on different days are independent random variables.

We have then the recursion

$$F_n(y) = \max_{N \geqslant 0} E[-A(y) - bN - d_N + (a+b)X + F_{n-1}(y + N - X)]$$

$$(n = 1, 2, \ldots), \qquad \textbf{12.3.4}$$

where the expectation is with respect to demand, X. 12.3.4 is established by considering income and expenditure for the following day: overnight storage costs etc. of $A(y)$, wholesalers' costs of $bN + d_n$ next morning, receipts $(a+b)X$ during the day; and in the evening the newsagent is left with a stock of $y + N - X$, and only $n-1$ days to run. We must average over X, which is unknown at the time the order N is placed. The optimal order to place will be the value $N_n(y)$ maximizing the expectation in **12.3.4**.

Equation **12.3.4** is known as the *dynamic programming equation*. In constructing it, we have utilized the fact that the actual stocking rule used when there are $1, 2, 3, \ldots$ days left to run need not be stated explicitly: all we need know is the function $F_{n-1}(y)$ determining future expected profits if the stock is y and there are $n-1$ days to run. The optimal order for the nth last day, $N_n(y)$, and the expected profit, $F_n(y)$, are then determined in terms of $F_{n-1}(y)$ by the recursion **12.3.4**. Our primary interest may be in the actual ordering policy as determined by the function $N_n(y)$, but in order to determine this, we must also bring in the profit function $F_n(y)$.

We have now to solve the functional equation **12.3.4** recursively, starting from $F_0(y) = 0$, and working through $n = 1, 2, \ldots$ consecutively, and determining the maximizing values $N_n(y)$ on the way. A mathematical solution would be the ideal, but these functional equations are often so difficult that we must resort to a numerical solution. However, a number of features of the solution can be determined by simple means.

For instance, we see fairly directly that the optimal procedure must be of the following form. There are two sequences of constants, $\{s_n\}$ and $\{S_n\}$, with $s_n < S_n$. If, with n days to go, y is greater than s_n, the newsagent should make no order at all. However, if y is less than s_n, he should bring stocks up to S_n by ordering an amount $S_n - y$. It is plausible that, if $n \to \infty$ (i.e. if we consider indefinite operation), then s_n and S_n tend to limit values s and S. This procedure is then known as an (s, S) *inventory policy*; its justification is as follows.

We see from **12.3.4** that, with stock y and n days to go, we must choose N as the positive integer maximizing $\psi_n(y + N) - d_n$, where

$$\psi_n(y + N) = E[F_{n-1}(y + N - X) - b(y + N - X)].$$ **12.3.5**

Now, the function $-A(y)$ is concave, i.e. $A(y)$ is convex, and we find by induction from **12.3.4** that the same is true of $F_n(y)$ and $\psi_n(z)$, as functions of y and z respectively. Suppose that $\psi_n(z)$ has its maximum (necessarily unique) at $z = S_n$, and let s_n be the smaller root of

$$\psi(z) = \psi(S_n) - d.$$ **12.3.6**

We see from Figure 9 that the procedure asserted is correct: the positive value maximizing $\psi_n(y + N) - d_N$ is

$$N_n(y) = \begin{cases} 0 & (y \geqslant s_n), \\ S_n - y & (y < s_n). \end{cases}$$ **12.3.7**

Further analysis is needed to determine the critical values s_n and S_n, but it is a great help that the character of the policy is determined.

This is a typical sequential decision problem; one can easily find similar problems in less domestic contexts, and these awaken increasing interest. We can mention control theory, missile guidance, production control, sequential experimentation, statistical inference, learning and self-adapting mechanisms.

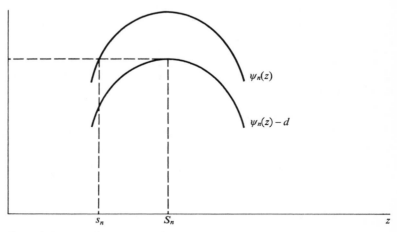

Figure 9 Determination of the critical values for stock control

One could give a rather general formulation of a sequential decision problem in discrete time. Suppose the decision to be taken at time t is the choice of value (not necessarily numerical) of a quantity u_t. In our example this was just the order size N. We wish to choose, the $\{u_t\}$ sequence so as to maximize $E(g)$, where the 'utility' g may be a function of many variables and r.v.s, including the u_t. (In the example g was

$$\sum [(a+b)X_t - A(y_t) - bN_t - d_{N_t}];$$

and in this formal description we shall not worry about the fact that this sum may be infinite.) Now, u_t can only be determined in terms of the quantities one knows at time t, the observations whose values are actually available then. Let these be collectively denoted by w_t. (In the example w_t is y_s, X_s and N_s ($s < t$), together with n_t, the number of days yet to run at time t.)

If $G(w_t)$ denotes the total expected maximal utility, conditional on w_t, then we have the *dynamic programming equation*

$$G(w_t) = \max_{u_t} E[G(w_{t+1})|w_t, u_t]. \qquad \textbf{12.3.8}$$

The maximizing value determines a function $u(w_t)$ describing the optimal policy.

Equation **12.3.8** is a good deal more complex than it appears, since w_t will in general be an infinite collection of variables. However, individual features of a process can lead to the simplification that optimal decision and expected 'residual utility' at time t depend upon a much smaller set of variables. In our case y_{t-1} and n_t were sufficient; essentially because we assumed the X_t to be independently distributed. (In fact this was taken for granted, without proof. A proof and a generalization are indicated in exercise 12.3.3.)

Exercises

12.3.1 Suppose that stock cannot be replenished from other sources, and that excess demand on any day must remain unsatisfied. Show then that equation **12.3.4** is modified by the substitution of $y_+ + N - X$ for $y + N - X$ in the argument of F_{n-1}, if $A(y)$ is considered to represent losses due to unsatisfied demand when y is negative. Show that this equation is susceptible to the same treatment as **12.3.4**.

12.3.2 Sometimes one works with *discounted future profits*, so that a pound tomorrow is considered to be worth only α pounds today $(0 < \alpha < 1)$ since an investment of α today will appreciate to unity tomorrow by compound interest. Show that the F_{n-1} in **12.3.4** should then be multiplied by α, and that one can then expect $F_n(y)$ to tend to a finite limit $F(y)$ as n increases. Show that the same qualitative conclusions on an optimal order policy hold.

12.3.3 Suppose that the sequence of demands $\{X_t\}$ is Markov. By starting from $F_0 = 0$, and using the general recursion **12.3.4**, show that at time t, and with n days to go, F is a function $F_n(y_{t-1}, X_{t-1})$ of n, y_{t-1} and X_{t-1}, obeying the recursion

$$F_n(y, X_{t-1})$$
$$= \max_{N \geqslant 0} E\left[-A(y) - bN - d_N + (a+b)X_t + F_{n-1}(y + N - X_t, X_t) \mid X_{t-1} \right].$$

One could say that the reason why X_{t-1} now appears in F and N is because it helps to predict future demand.

12.4 Stochastic differential equations; generalized processes

We have from time to time considered the stochastic difference equation

$$X_t - \rho X_{t-1} = \varepsilon_t, \qquad \qquad \textbf{12.4.1}$$

where 'time' t takes integral values, $\{\varepsilon_t\}$ is a sequence of r.v.s, and $\{X_t\}$ a sequence derived from it by use of the recursion **12.4.1**. Models such as **12.4.1** or higher order versions of it, occur repeatedly in physical applications. For example, X_t might be the amount of water in a lake in the tth year, ρX_{t-1} the amount retained from the previous year, and ε_t the inflow since the previous year.

One might postulate all kinds of statistical properties for the ε_t variables; the simplest would be to assume them uncorrelated, with constant mean μ and constant variance σ^2. This would not be very realistic for the lake model, but might be an acceptable first approximation.

A convenient way of stating this assumption is to require that

$$E\left(\sum \theta_t \varepsilon_t\right) = \mu \sum_t \theta_t, \qquad\qquad \textbf{12.4.2}$$

$$\operatorname{var}\left(\sum \theta_t \varepsilon_t\right) = \sigma^2 \sum_t \theta_t^2, \qquad\qquad \textbf{12.4.3}$$

for all sequences of real constants $\{\theta_t\}$ for which the two series are absolutely convergent. In other words, one postulates the expectation values of the r.v.s

$$\sum \theta_t \varepsilon_t \quad \text{and} \quad \left(\sum \theta_t \varepsilon_t\right)^2$$

This has advantages when we come to processes which are more difficult to specify in terms of distributions of individual r.v.s etc. Also, in the present case this formulation is a natural one, because the solution of **12.4.1** in terms of past inflows is

$$X_t = \sum_0^\infty \rho^s \varepsilon_{t-s}, \qquad\qquad \textbf{12.4.4}$$

which is just a linear form such as $\sum \theta_t \varepsilon_t$.

We can say immediately from **12.4.2** and **12.4.3** that

$$E(X_t) = \mu \sum_0^\infty \rho^s = \frac{\mu}{1-\rho}, \qquad\qquad \textbf{12.4.5}$$

$$\operatorname{var}(X_t) = \sigma^2 \sum_0^\infty \rho^{2s} = \frac{\sigma^2}{1-\rho^2}, \qquad\qquad \textbf{12.4.6}$$

these expressions being convergent if $|\rho| < 1$ (the criterion for *stability* of the system, and also for m.s. convergence of the sum **12.4.4**: see section 9.7).

In fact, we see from **12.4.3** and **12.4.4** that

$$\operatorname{var}\left(\sum \theta_t X_t\right) = \operatorname{var}\left(\sum_t \sum_{s \geq 0} \theta_t \rho^s \varepsilon_{t-s}\right)$$

$$= \sigma^2 \sum_t (\theta_t + \rho\theta_{t+1} + \rho^2\theta_{t+2} + \ldots)^2$$

$$= \sigma^2 \sum_s \sum_t \theta_s \theta_t \frac{\rho^{|s-t|}}{1-\rho^2}, \qquad\qquad \textbf{12.4.7}$$

so that $\quad \operatorname{cov}(X_s, X_t) = \dfrac{\sigma^2 \rho^{|s-t|}}{1-\rho^2}. \qquad\qquad \textbf{12.4.8}$

These methods are also applicable to complete distributions: we might formulate the assumption that the ε_t are jointly normally distributed with the previous first and second order moments by specifying the expectations

$$E(e^{i\Sigma \theta_t \varepsilon_t}) = \exp\left(i\mu \sum \theta_t - \tfrac{1}{2}\sigma^2 \sum \theta_t^2\right) \qquad\qquad \textbf{12.4.9}$$

for the same real θ_t as before. From **12.4.4** and **12.4.9** we then find that

$$E(e^{i\Sigma\theta_t X_t}) = \exp\left(\frac{i\mu}{1-\alpha}\sum\theta_t - \frac{\sigma^2}{2(1-\rho^2)}\sum_s\sum_t \rho^{|s-t|}\theta_s\theta_t\right) \qquad \textbf{12.4.10}$$

That is, the X_t are also jointly normally distributed with first and second moments **12.4.5** and **12.4.8**.

The specification of the process in terms of the expectations **12.4.9** (effectively, the joint c.f. of an infinity of r.v.s) means that only certain relatively smooth functions of the sequence $\{\varepsilon_t\}$ can be studied (see section 11.2), but such functions are probably just those of physical interest. In particular, the convergent linear functions $\sum\theta_t\varepsilon_t$ are easily studied, and these, in view of **12.4.4**, are of immediate physical interest.

These considerations extend immediately to continuous time processes: consider the first-order stochastic differential equation analogous to **12.4.1**:

$$\frac{dX(t)}{dt} + \alpha X(t) = \varepsilon(t). \qquad \textbf{12.4.11}$$

For example, $X(t)$ might be the velocity of a Brownian particle at time t, and $\varepsilon(t)$ represent the acceleration caused by random impacts on the particle. The term $\alpha X(t)$ represents a damping or braking effect (if $\alpha > 0$), due to the fact that the particle will receive more collisions from the direction in which it is moving than from the opposite direction, this effect increasing with velocity.

The solution of **12.4.11** in terms of past impulses, corresponding to **12.4.4**, is

$$X(t) = \int_0^\infty e^{-\alpha s}\varepsilon(t-s)\,ds. \qquad \textbf{12.4.12}$$

Suppose we now postulate for the $\{\varepsilon(t)\}$ process that

$$E\left[\int\theta(t)\varepsilon(t)\,dt\right] = 0, \qquad \textbf{12.4.13}$$

$$E\left[\left\{\int\theta(t)\varepsilon(t)\,dt\right\}^2\right] = \int\int K(s-t)\theta(s)\theta(t)\,ds\,dt, \qquad \textbf{12.4.14}$$

which is equivalent to saying that $\varepsilon(t)$ has zero mean, and

$$\mathrm{cov}[\varepsilon(s),\,\varepsilon(t)] = K(s-t). \qquad \textbf{12.4.15}$$

This is in fact a self-consistent specification if K is such that expression **12.4.14** is never negative. We see from **12.4.12–14**, as before, that $X(t)$ has zero mean and

$$\mathrm{cov}[X(s),\,X(t)] = \int_0^\infty du \int_0^\infty dv\, e^{-\alpha(u+v)} K(s-t-u+v). \qquad \textbf{12.4.16}$$

229 **Stochastic differential equations; generalized processes**

Now, suppose we wish to assume, in analogy with our first process **12.4.1**, that $\varepsilon(s)$ and $\varepsilon(t)$ are uncorrelated for $s \neq t$, corresponding to the assumption that the impulses the Brownian particle receives from surrounding molecules are 'completely irregular' in some sense. In analogy with **12.4.3** we should presumably then require that

$$E\left[\left\{\int \theta(t)\varepsilon(t)\,dt\right\}^2\right] = \sigma^2 \int \theta(t)^2\,dt \qquad\qquad \textbf{12.4.17}$$

for all functions $\theta(t)$ for which the integral is meaningful and convergent. However, upon comparison with **12.4.14** and **12.4.15** we see that **12.4.17** implies

$$\mathrm{var}[\varepsilon(t)] = \infty \quad \text{if} \quad \sigma^2 > 0.$$

There is thus a dilemma. If one is to have an 'uncorrelated' process whose integral is to have positive variance, then the process variable itself must have infinite variance. If the Brownian particle is to receive uncorrelated accelerations which have a net effect, then the acceleration must sometimes be infinite, i.e. impulsive.

For a long time such a process was regarded as 'improper', although it was nevertheless used, formally, as a convenient ideal ('white noise') which can be approximated arbitrarily closely by 'proper' processes (cf. exercise 12.4.6), and which is indeed approximated by physical processes (e.g. 'thermal noise' in electronic circuits). However, the concept can be used freely and consistently, in that one may specify the property **12.4.17** and appeal to this where necessary.

The situation is closely analogous to the Dirac δ-function, which has the 'property'

$$\int \delta(t)\theta(t)\,dt = \theta(0) \qquad\qquad \textbf{12.4.18}$$

for all suitable functions $\theta(t)$. This function was similarly regarded as improper, but also used formally, as a convenient ideal which could be approximated arbitrarily closely by proper functions. However, one never uses a δ-function in isolation, but only in the form of an integral such as **12.4.18**. So, if one specifies the δ-function as having the property **12.4.18**, then one has the property one needs operationally, and can use it consistently. Development of this point of view has led to the theory of *generalized functions*.

Similarly, one never uses the 'white noise' process in isolation, but only in integrals such as $\int \theta(t)\varepsilon(t)\,dt$. The prescription **12.4.17** is self-consistent, and exactly what one needs for dealing with the second order properties of linear functions of the process. In this way one arrives at a theory of *generalized processes*, by specifying in a self-consistent fashion the expectations of just those r.v.s which are physically meaningful.

Of course, one need not restrict attention to second order properties, but can specify, for example, the continuous time analogue of the characteristic function **12.4.9** (see exercise 12.4.6).

Exercises

12.4.1   Show that, under assumption **12.4.17**, the integral **12.4.12** is m.s. convergent if $\alpha > 0$.

12.4.2   Evaluate **12.4.16** in the case **12.4.17**.

12.4.3   Suppose one replaces **12.4.11** by

$$\frac{dX}{dt} + \alpha X + \beta Y = \varepsilon,$$

where Y is the coordinate of the Brownian particle (so that $X = \dot{Y}$), and assumes that

$$E\left[\exp\left\{ i \int \theta(t)\varepsilon(t)\, dt \right\} \right] = \exp\left\{ -\tfrac{1}{2}\sigma^2 \int \theta^2(t)\, dt \right\}.$$

Determine the joint distribution of $X(t)$ and $Y(t)$ in equilibrium.

12.4.4   Suppose that

$$\varepsilon(t) = \sum_j \xi(t - t_j),$$

where the t_j are the instants at which an event takes place in a Poisson process of intensity λ, and ξ is a given function. Thus, with each event in a Poisson process there is associated a 'pulse', having value $\xi(s)$ a time s after occurrence of the event. Show that formula **7.5.11** may be generalized to

$$\Phi(\theta) = E[\exp\{i \int \theta(t)\varepsilon(t)\, dt\}] = \exp\left[\lambda \int \left\{ \psi(t) - 1 \right\} dt \right], \qquad \textbf{12.4.19}$$

where $\psi(t) = \exp\left\{ i \int \theta(s + t)\xi(s)\, ds \right\}.$

12.4.5   Suppose the pulses of exercise 12.4.4 are random in shape, so that

$$\varepsilon(t) = \sum_j \xi_j(t - t_j),$$

where the ξ_j are independently and identically distributed random functions, independent of the t_j. Show that **12.4.19** holds, with

$$\psi(t) = E\left[\exp\left\{ i \int \theta(s + t)\xi(s)\, ds \right\} \right] \qquad \textbf{12.4.20}$$

12.4.6   Show from **12.4.19** and **12.4.20** that, if $E(\xi) = 0$, then

$$\text{var}\left[\int \theta(t)\varepsilon(t)\,dt \right] = \lambda \int \int \int E[\xi(s)\xi(u)]\theta(s+t)\theta(u+t)\,dt\,ds\,du. \qquad \textbf{12.4.21}$$

We leave the reader to convince himself heuristically that, if the pulses are made short enough and strong enough, then expression **12.4.21** will approximate to

$$\text{constant} \times \int \theta(t)^2\,dt\,;$$

also that if they are made frequent enough, by increase of λ, then expression **12.4.19** will approximate to

$$\exp\left[-\text{constant} \times \int \theta(t)^2\,dt \right]$$

Both these conclusions demand some degree of continuity of $\theta(t)$.

References

APOSTOL, T. M. (1957), *Mathematical Analysis*, Addison-Wesley.

FEINSTEIN, A. (1958), *Foundations of Information Theory*, McGraw-Hill.

FELLER, W. (1966), *An Introduction to Probability Theory and its Applications*, Vol. II, Wiley.

GOLDBERG, R. R. (1961), *Fourier Transforms*, Cambridge University Press.

HARRIS, T. E. (1963), *The Theory of Branching Processes*, Springer; Prentice-Hall.

KARLIN, S. (1959), *Mathematical Methods and Theory in Games, Programming and Economics*, Vol. I, Pergamon; Addison-Wesley.

KERRICH, J. E. (1946), *An Experimental Introduction to the Theory of Probability*, Einar Munksgaard, Copenhagen.

KINGMAN, J. F. C., and TAYLOR, S. J. (1966), *Introduction to Measure and Probability*, Cambridge University Press.

KOLMOGOROV, A. N. (1933), *Grundbegriffe der Wahrscheinlichkeitsrechnung*, Ergebnisse der Math., English edition, Chelsea, 1950.

KRICKEBERG, K. (1965), *Probability Theory*, Addison-Wesley.

NEVEU, J. (1964), *Bases Mathématiques du Calcul des Probabilités*, Masson, Paris; English edition, Holden-Day, 1965.

SHANNON, C. E., and WEAVER, W. (1949), *The Mathematical Theory of Communication*, University of Illinois Press.

Index